Progress in Molecular and Subcellular Biology 17

Series Editors

Ph. Jeanteur, Y. Kuchino,
W.E.G. Müller (*Managing Editor*)
P.L. Paine

Springer
Berlin
Heidelberg
New York
Barcelona
Budapest
Hong Kong
London
Milan
Paris
Santa Clara
Singapore
Tokyo

G. Csaba W.E.G. Müller (Eds.)

Signaling Mechanisms in Protozoa and Invertebrates

With 70 Figures

Springer

Prof. Dr. G. CSABA
Department of Biology
Semmelweis University of Medicine
P.O. Box 370
Nagyvárad tér 4
1445 Budapest
Hungary

Prof. Dr. W.E.G. MÜLLER
Institut für Physiologische Chemie
Abteilung Angewandte Molekularbiologie
Universität Mainz
Duesbergweg 6
55099 Mainz
Germany

ISBN-13: 978-3-642-80108-2 e-ISBN-13: 978-3-642-80106-8
DOI: 10.1007/978-3-642-80106-8

Library of Congress Cataloging-in-Publication Data. Signaling mechanisms in protozoa and invertebrates/G. Csaba, W.E.G. Müller. p. cm. – (Progress in molecular and subcellular biology: 17) Includes bibliographical references and index. ISBN 3-540-60796-X 1. Protozoa – Physiology. 2. Invertebrates – Physiology. 3. Hormone receptors. 4. Cellular signal transduction. 5. Endocrinology, Comparative. I. Csaba, György. II. Müller, W.E.G. (Werner E.G.). 1942– III. Series. QH506. P76 no. 17 [QL369.2] 574.8'8 s – dc20 [593.1'041] 96-3924

Softcover reprint of the hardcover 1st edition 1996

Cover design: Design & Production, Heidelberg

SPIN: 10499340 39/3137/SPS – 5 4 3 2 1 0 – Printed on acid-free paper

Preface

Comparative endocrinology helps to find the roots of homeostatic regulation in organisms. In this context, many years ago a series of experiments were done, which demonstrated the hormonal regulation also on the invertebrate level. The mechanisms are partly similar, partly different, from those found in vertebrates. The new receptor era of mammalian endocrinology stimulated research on invertebrate hormone receptors, and sophisticated methods are applied also to determine hormones. The experiments demonstrated the existence and even similar function of these structures and signaling molecules. However, data on hormones and receptors at the lowest level of metazoan life and the highest level of protozoan life were not at our disposal.

About two decades ago, first observations on the presence of hormone receptors reacting to vertebrate hormones in protozoa were made. Since the early 1980s we know that hormone-like molecules similar to those of higher vertebrates are present also in unicellular organisms. The presence of some second messengers in *Tetrahymena* was recognized. Since then, the research has been extended and many structures – previously believed to be solely vertebrate characteristics, such as opiate receptors, similar to mammalian ones – were found in unicellular organisms. These observations justified the assumption of a complete endocrine system at protozoan level, where – considering the unicellularity – this seemed to be not required. However, it became clear that the roots of endocrine communication date back at least 2 billion years.

Other experiments proved the existence of hormones and genes of hormone receptors in animals at the lowest level of metazoan life with very high homology to mammalian and vertebrate hormones. Many members of different recognition systems have already been studied. Basing on these investigations, the universality of biological mechanisms seemed to be further clarified, and new light was shed on the evolution of recognition (receptor-signal) systems. This was supported by observations (e.g., on pheromones, chemokinesis, etc.) which are more characteristic for the lower level of phylogeny than for the vertebrates.

In a world of cellular communication, mechanisms other than hormonal systems also have a very important role. Recognition and signal molecules, such as immunoglobulins and lectins, can transmit information in animal organisms, including invertebrates. The study of these problems and their interrelations with endocrine regulation help to form a coherent picture of the universalities of cell-to-cell communication in the living world.

Nevertheless, the thorough study of signaling mechanisms is important not only from a theoretical point of view, but as providing researchers with a tool to study vertebrate – mammalian and human – endocrinology. Due to the similarities in the participating structures, molecules, and mechanisms of mammalian and invertebrate (protozoan) endocrine communication, these events can be studied more easily in the more simple organisms, the lower invertebrates.

Many of the authors of this book were (are) the initiators, others the continuers of the work mentioned. We believe that the information, compiled by them in the book, will promote further success in this field of research.

Budapest and Mainz, March 1996

G. Csaba
W.E.G. Müller

Contents

List of Contributors

Addresses are given at the beginning of the respective contribution

Ahmed, H. 158
Chiesa, R. 29
Csaba, G. 1
Grave, M. 61
Hellung-Larsen, P. 61
Kovács, P. 105
Leick, V. 61
Lipke, P.N. 119
Luporini, P. 80
Marino, M. 29
Miceli, C. 80
Müller, W.E.G. 183
Nozawa, Y. 40
Ortenzi, C. 80
Renaud, F.L. 29
Rodríguez, F. 29
Schäcke, H. 183
Tomassini, N. 29
Umeki, S. 40
Vallesi, A. 80
Vasta, G.R. 158

Evolutionary Significance of the Hormone Recognition Capacity in Unicellular Organisms. Development of Hormone Receptors

G. CSABA[1]

1 Introduction

Protozoa live in water, where they are surrounded by a mass of molecules that affect their life. Among these molecules, food molecules are considered as useful, yet some others are noxious or toxic. Consequently, the recognition and discrimination of molecules, by the protozoa, is not only essential but crucial.

The unicellular ciliate *Tetrahymena* can recognize different hormones characteristic of vertebrate animals. It can respond to "classical" hormones of amino acid type such as thyroxin (Csaba and Németh 1980), epinephrine (Csaba and Lantos 1976), and serotonin (Csaba and Lantos 1973), to polypeptide hormones such as insulin (Csaba and Lantos 1975), vasopressin, and oxytocin (Csaba and Kovács 1992), or the hormones of the pituitary gland (Csaba and Ubornyák 1982a), and to the recently discovered hormones such as epidermal growth factor (Csaba and Kovács 1991), endothelin (Köhidai and Csaba 1995), atrial natriuretic peptide (Köhidai et al. 1995), or cytokines (Csaba et al. 1995). It can also respond to plant hormones, such as indoleacetic acid (Csaba and Ubornyák 1982b; Csaba and Darvas 1984). It also has drug receptors, e.g., for opiates (O'Neil et al. 1988; Zipser et al. 1988) or benzodiazepines (Csaba et al. 1989).

Tetrahymena can differentiate between the related hormones of a family, or between the functional hormones and their precursors. Serotonin can promote the phagocytosis of *Tetrahymena*; however, its close relative indolacetic acid cannot (Csaba and Lantos 1973). Vasopressin and oxytocin – differing only in two amino acids – disparately influence the contractile vacuole of *Tetrahymena* (Csaba and Kovács 1992). Thyroxin and its precursors triiodothyronine, diiodotyrosine (T2), and monoiodotyrosine differently influence the growth of *Tetrahymena*, T2 most strongly (Csaba and Németh 1980).

These facts demonstrate that the unicellular *Tetrahymena* has a wide recognition capacity in accordance with its milieu. In this respect, the abovementioned hormones could be simply molecules recognized by the protozoa without being necessarily "hormones", in the strict sense, for them. However,

[1]Department of Biology, Semmelweis University of Medicine, P.O. Box 370, 1445 Budapest, Hungary

these molecules provoke a response, which in many cases is characteristic of the vertebrates. Insulin influences sugar metabolism (Fig. 1), as does epinephrine (Csaba and Lantos 1975, 1976). Histamine increases phagocytosis (Csaba and Lantos 1973), vasopressin influences water turnover (Csaba and Kovács 1992), and thyroxin stimulates cell multiplication (Csaba and Németh 1980). This means that a hormone is not simply "one of the surrounding molecules" for *Tetrahymena,* but a signal molecule whose message could be perceived. Based on these observations, the second messenger system of *Tetrahymena* has also been discovered (Csaba and Nagy 1976; Csaba et al. 1976, 1987; Kuno et al. 1979; Kovács and Csaba 1987, 1990a).

The receptor of *Tetrahymena* seems to be similar to the receptors of higher-ranked animals. The opiate receptor has been studied and sequenced, and it was found to be similar to those of leech and rat (Zipser et al. 1988).

It is not easy to explain how *Tetrahymena* is able to recognize all the previously unknown molecules present around it. This requires a very dynamic membrane with unfixed protein structures moving continuously and building and destroying different membrane patterns (Koch et al. 1979). In fact, this is not impossible if we consider the fluid mosaic model for the membrane. Considering this, the receptor of *Tetrahymena* is nothing but a suitable membrane pattern which is accidentally present when the unicellular organism encounters the given molecule. If we suppose that the protein's movement and the pattern formation in the plasma membrane is very fast, and knowing that the variations are unlimited, an encounter between the "receptor" and "signal molecule" is not unlikely.

It is worth mentioning that not only *Tetrahymena* but other unicellular organisms, such as *Blepharisma undulans* (Kovács and Csaba 1990b), *Amoeba proteus* (Köhidai et al. 1992; Csaba et al. 1984c), and fungi (McKenzie et al. 1988), even *Acetabularia* (Legros et al. 1975), have hormone receptors. Nevertheless, the best model for these studies is *Tetrahymena* (Csaba 1981, 1985, 1994).

2 Receptor Memory: Hormonal Imprinting

Exposure to the presence of a noxious substance in the surroundings could be detrimental for the Protozoa. The cell which senses the presence of the sub-

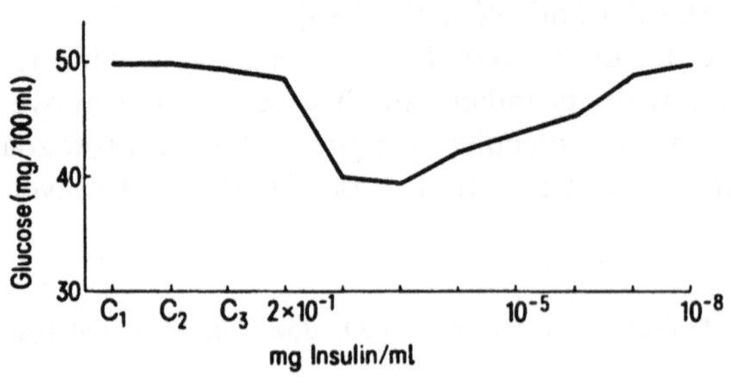

Fig. 1. Effect of insulin on the glucose uptake of *Tetrahymena pyriformis.* C_1 Glucose control; C_2 glucose + insulin control; C_3 glucose + *Tetrahymena* control. At high concentrations of insulin the glucose uptake increases significantly. (Csaba and Lantos 1975)

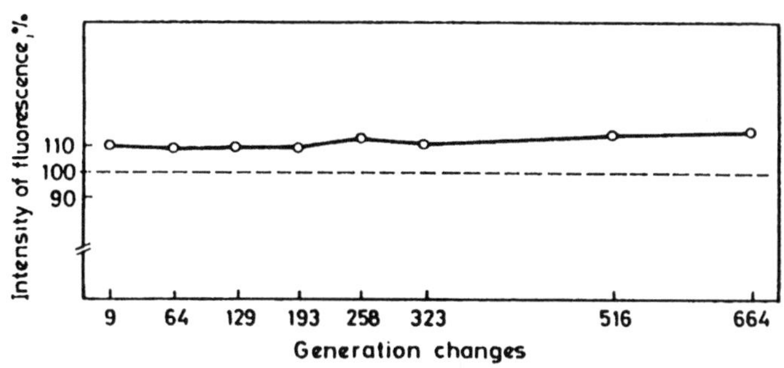

Fig. 2. FITC-insulin binding in the offspring generations of *Tetrahymena* cells treated with hormone on a single occasion related to the control as 100%. The insulin binding is always higher and does not decrease with the progressing number of cell divisions. (Köhidai et al. 1990)

stance in a very low concentration escapes and survives, others die. Similarly, the cell which recognizes the presence of the food earlier, will eat more and survive. However, the life span of the individual *Tetrahymena* is very short. This means that the information gained has to be somehow transmitted to the progeny generations, to make recognition easier and help the survival of species.

After the first encounter, the *Tetrahymena* "remembers" the hormone, and this memory is transmitted to the progeny. After about 600 generations, the cells still remember (Fig. 2) and bind the hormone with higher affinity and respond to it differently from how they did in the case of the first encounter. This phenomenon is named hormonal imprinting (Csaba 1980).

3 Problems of the Specificity of Imprinting

Hormonal imprinting – the first meeting of the to-be hormones and the to-be receptors – specifies the membrane pattern, or increases their number in the plasma membrane. Studying hormone binding there is always some – high – aspecific binding, possibly due to the glycoproteins or glycosyl phosphatidyl inositols on the cell's surface. However, 24 h after imprinting – in the progeny generations – the specific binding significantly increases (Kovács and Csaba 1990c; Figs. 3, 4). The same was demonstrated using confocal microscopy (Christopher and Sunderman 1992).

The binding site specifying the effect of imprinting is demonstrable by using direct or indirect methods. *Tetrahymena* cells treated with antibodies to rat liver cell membrane (insulin receptor) exhibit a considerable increase in binding capacity on reexposure to the antibody 24 h later (Fig. 5). Insulin binding is similarly enhanced by preexposure to the antibody and, conversely, preexposure of *Tetrahymena* to insulin increases the later binding of rat liver receptor antibodies (Csaba et al. 1984b). This means that there is a complementarity between the receptor developed by imprinting and the imprinting-

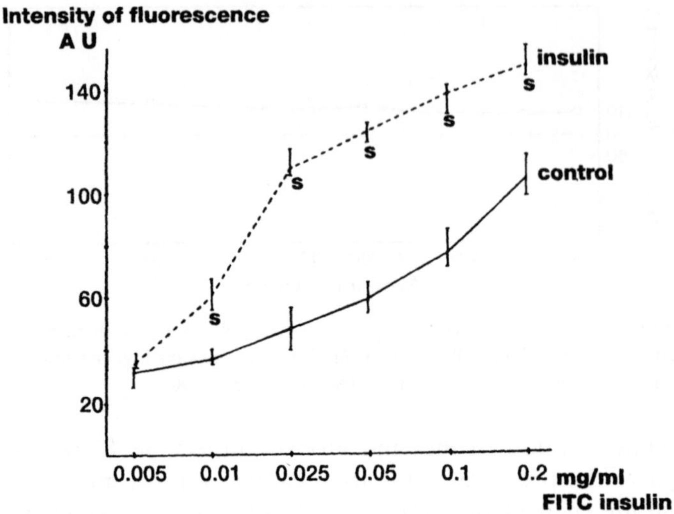

Fig. 3. Saturation of binding sites (receptors) of *Tetrahymena* after 30-min incubation with increasing FITC-insulin concentration in untreated (control) and insulin-imprinted (insulin) *Tetrahymena* cells. $s = p < 0.01$ to control. The imprinted cells bind more insulin. (Kovács and Csaba 1990c)

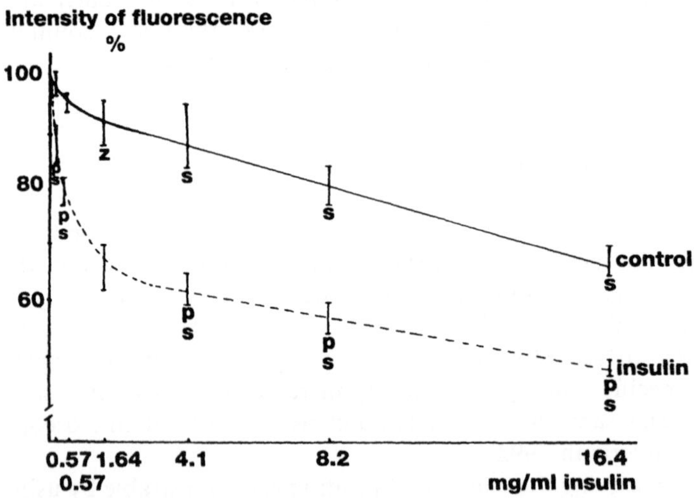

Fig. 4. Displacement of FITC-insulin by unconjugated insulin in control and insulin imprinted (insulin) *Tetrahymena* cells. $p = p < 0.01$ to the control; $s = p < 0.01$; $z = p < 0.05$ to the 100% (cells not treated with unconjugated insulin). There is enhanced displacement after insulin imprinting, which points to the specificity of receptors. (Kovács and Csaba 1990c)

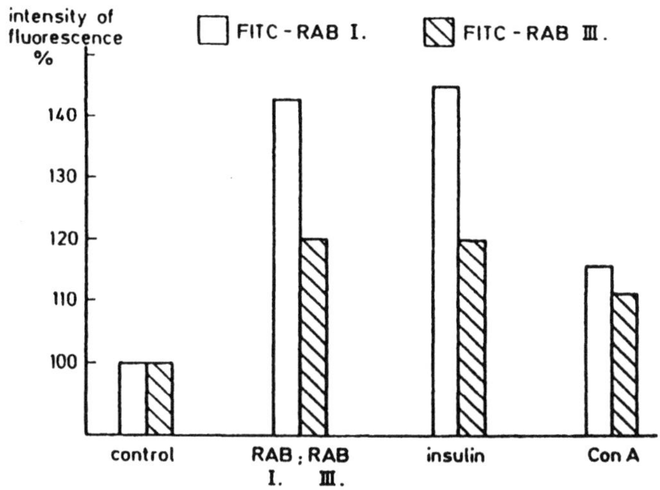

Fig. 5. Binding of FITC-labeled liver insulin receptor antibodies (*RAB I* and *RAB III*) to control and (*RAB*, *insulin*, or *Con-A*) pretreated *Tetrahymena* cells (*abscissa*). $p < 0.01$, except Con-A. The specific antibody treatment provokes the increase of the capacity of binding sites able to bind insulin and RAB alike. (Csaba et al. 1984b)

provoking ligand, and demonstrates that *Tetrahymena* and rat receptors are essentially similar. Another indirect confirmation of receptor specification and similarity to vertebrate receptors was available when antibodies were produced in rabbits against normal and insulin-imprinted *Tetrahymena* (Kovács et al. 1985). The antibody directed towards insulin-treated *Tetrahymena* showed significantly higher affinity to rat liver insulin receptors (Figs. 6, 7). The

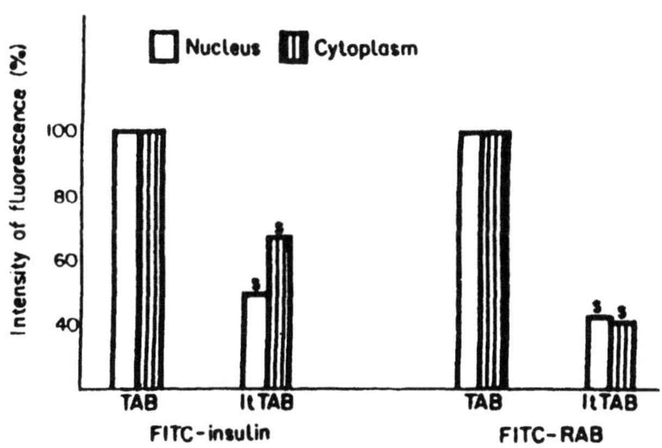

Fig. 6. Binding of FITC-insulin and FITC-RAB (see Fig. 5) to rat liver cells treated (after fixation) with rabbit IgG to intact (*TAB*) and insulin-pretreated *Tetrahymena* cells (*ItTAB*). s = significance between TAB and ItTAB = $p < 0.01$. The antibody to imprinted *Tetrahymena* inhibited the binding of insulin or RAB related to the effect of antibody to untreated *Tetrahymena* (Kovács et al. 1985)

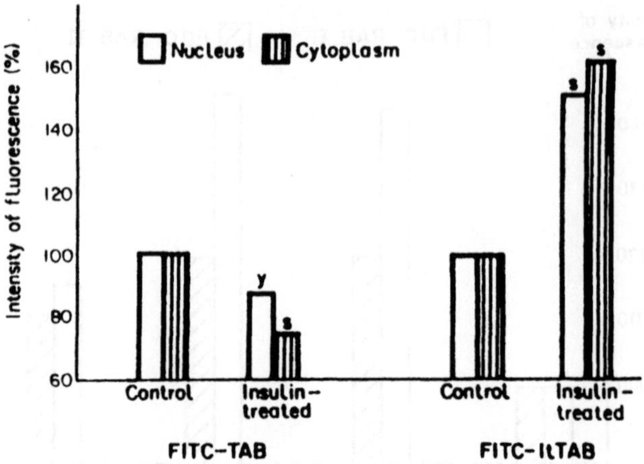

Fig. 7. Binding of FITC-TAB or FITC-ItTAB (see Fig. 6) to rat hepatocytes treated with insulin in vivo before fixation. $s = p < 0.01$; $y = p < 0.05$ to control. The binding of antibody to imprinted *Tetrahymena* was more expressed. The experiments, as well as those demonstrated in Fig. 6, show the receptor specifying effect of hormonal imprinting. (Kovács et al. 1985)

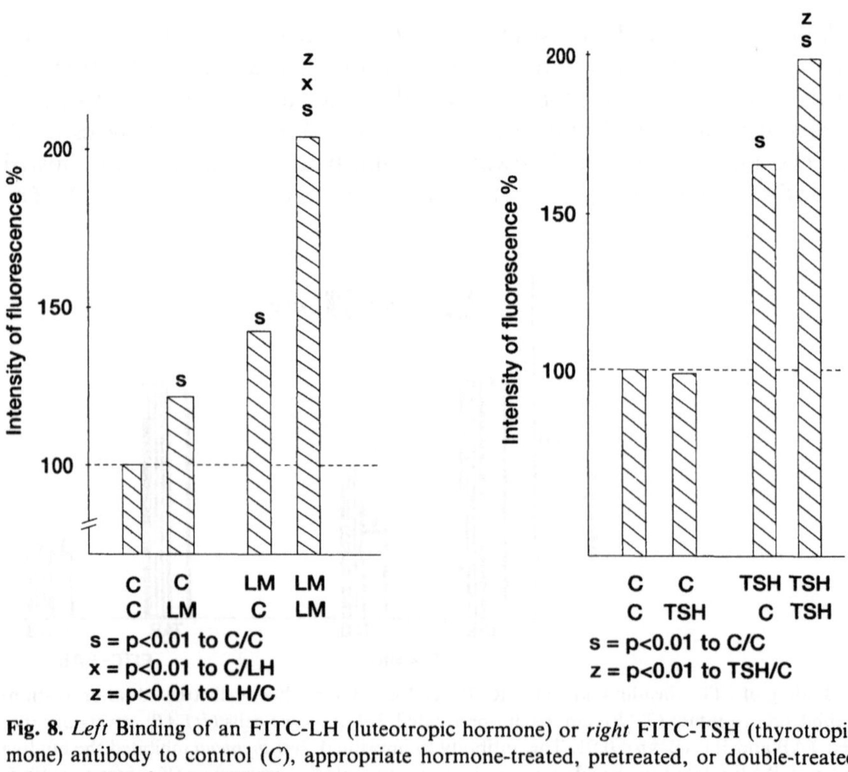

Fig. 8. *Left* Binding of an FITC-LH (luteotropic hormone) or *right* FITC-TSH (thyrotropic hormone) antibody to control (*C*), appropriate hormone-treated, pretreated, or double-treated *Tetrahymena* (Kovács et al. 1989)

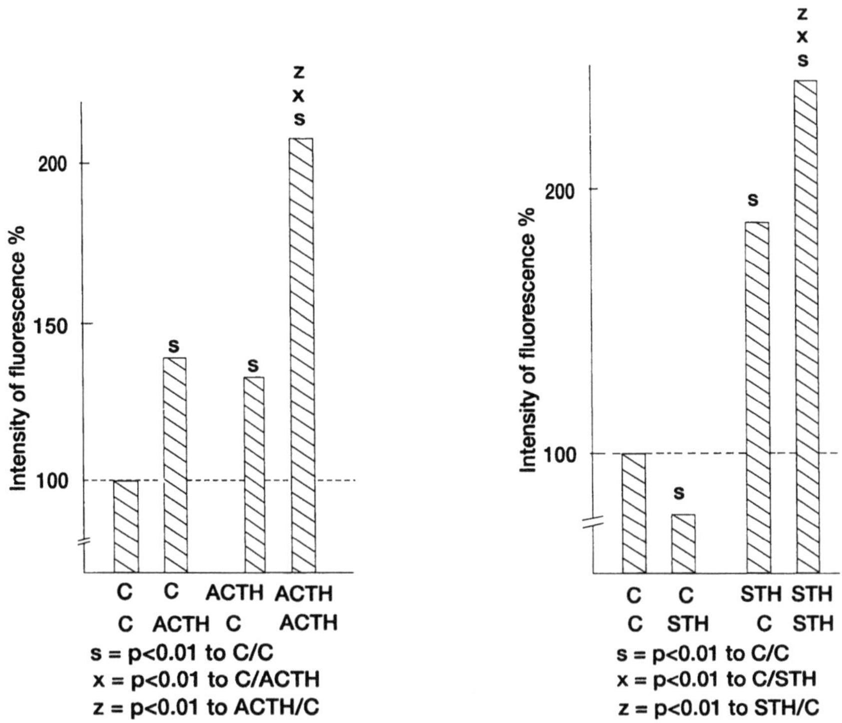

Fig. 9. *Left* Binding of an FITC-ACTH (adrenocorticotropic hormone) or *right* FITC-STH (somatotropic hormone) antibody to control (*C*), appropriate hormone-treated, pretreated, or double-treated *Tetrahymena*. The experiments, together with those demonstrated in Fig. 8, call attention to the receptor evoking (specifying) effect of the first encounter with the hormone. (Kovács et al. 1989)

binding promoting effect of imprinting has also been demonstrated by fluorescent labeled antibodies (Csaba and Hegyesi 1994).

Imprinting is specific for the evoking hormone (Figs. 8, 9). Insulin imprints for insulin, serotonin for serotonin, and so on (Kovács et al. 1989). Nevertheless, there is some overlap between the binding of related hormones. Thyrotropic hormone (TSH) develops imprinting for itself and for the follicle-stimulating hormone (FSH), and vice versa, probably due to the fact that both have identical alpha subunits and very similar beta subunits (Azukizawa et al. 1977; Kovács and Csaba 1985). However, the overlap is not larger than it used to be in the case of mammalian cells, and always less than the imprinting for themselves.

Not only hormones can evoke imprinting, but other molecules such as oligo- and polypeptides. However, the intensity and duration are different, depending on the structure of the molecule (Csaba et al. 1985). This could play a significant role in hormone and receptor evolution.

4 Time, Concentration, and Downregulation

In addition to the presence of hormone, time is also needed for the establishment of durable imprinting. Ten minutes' treatment of *Tetrahymena* does not provoke long-lasting effects, and only a weak effect is present after 1 h treatment. At least 4 h are needed for optimal, durable imprinting (Csaba et al. 1982). This supports the theory that the presence of changing membrane patterns is, in fact, also necessary for recognition.

Imprinting is not only time-, but also concentration-dependent (Csaba et al. 1982). T2 treatment in 10^{-9} M concentration is needed for 1 h to provoke optimal imprinting. A concentration of 10^{-18} M does not provoke imprinting or any response of the cell. However, after imprinting with 10^{-9} M T2, 10^{-18} M T2 can also provoke response, similar to that provoked by 10^{-9} M T2 on the first occasion (Table 1). Similarly, after imprinting with 10^{-18} M T2, repeated treatment with the same concentration evokes a response.

In the case of imprinting with a polypeptide hormone, at first, downregulation appears similar to that of mammals. Up to 6 h after a 1-h period of insulin treatment, there is no increase in insulin binding; this appears only after 24 h, and is even higher at 48 h (Figs. 10, 11). However, for downregulation, at least 3 h of treatment are needed (Csaba and Köhidai 1986). This also supports our view on the presence of changing membrane patterns, surmising that the proteins in the membrane are incessantly scanning the environment for information, and a certain time is required for the assembly of configurations capable of binding and provoking response.

Fourfold repetition of the 1-h treatment results in a more intensive response than a single treatment for 4 h (Csaba et al. 1984d). This supports the idea that imprinting is a memory-like process, also does the fact that many substances can weaken imprinting if given together with the main imprinting molecule, or even by the addition of new but different substances immediately after imprinting (Csaba et al. 1983b). However, on the addition of more than one

Table 1. Mean growth response of *Tetrahymena* to a single treatment with different concentrations of T_2

Hormone conc. (M)	10^{-9}	10^{-12}	10^{-15}	10^{-18}	Control
Growth rate in 24 h	21.0	17.6	14.6	12.2	12.4
Mean growth response of *Tetrahymena* to a single treatment with different concentrations of T_2 one week after 10^{-9} MT_2 pretreatment					
Growth rate in 24 h	20.7	18.6	19.6	19.6	12.2
Mean growth response of *Tetrahymena* to application of different hormone levels for first and second exposure					
First exposition as above reexposed at					
10^{-9} M	21.2	19.7	19.2	18.1	12.4
10^{-18} M	19.9	18.8	17.6	16.3	12.4

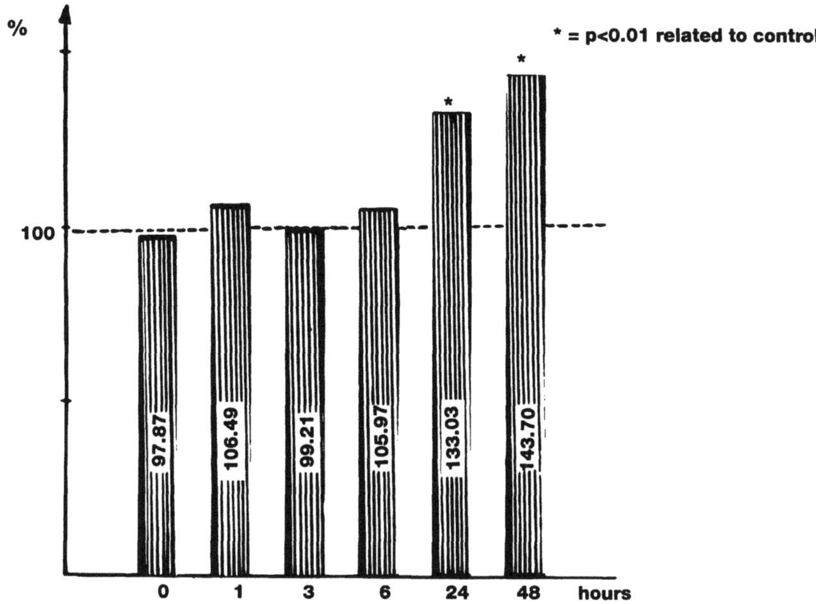

Fig. 10. Effect of 1-h insulin pretreatment (imprinting) on the binding of FITC-labeled insulin to *Tetrahymena* after maintenance for different periods in normal (insulin-free) medium. The effect of imprinting appears after 24 h. (Csaba and Köhidai 1986)

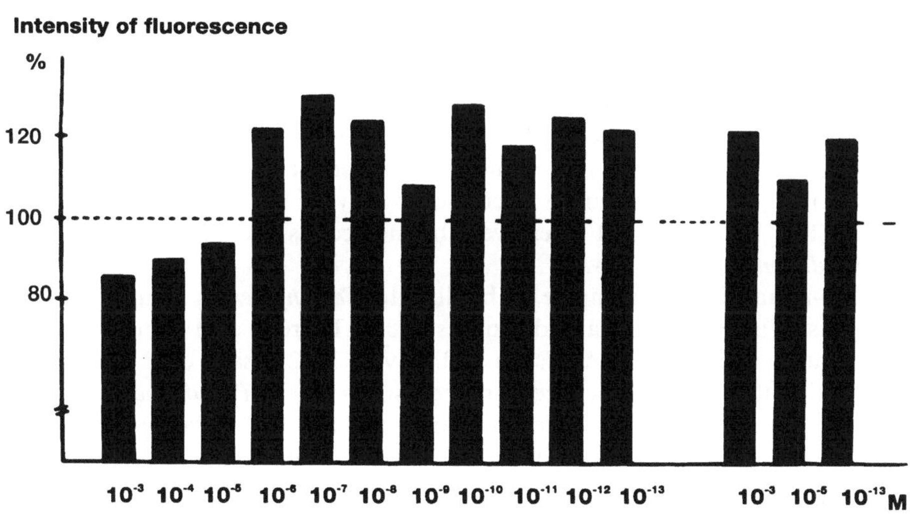

Fig. 11. Binding of FITC-insulin to *Tetrahymena* imprinted by different concentrations of insulin 1 day (*left side*) or 1 week (*right side*) after treatment, relative to the control as 100%. There is a down regulation by high concentrations after 24 h; however, these concentrations also caused imprinting, studied after 1 week. (Csaba and Köhidai 1986)

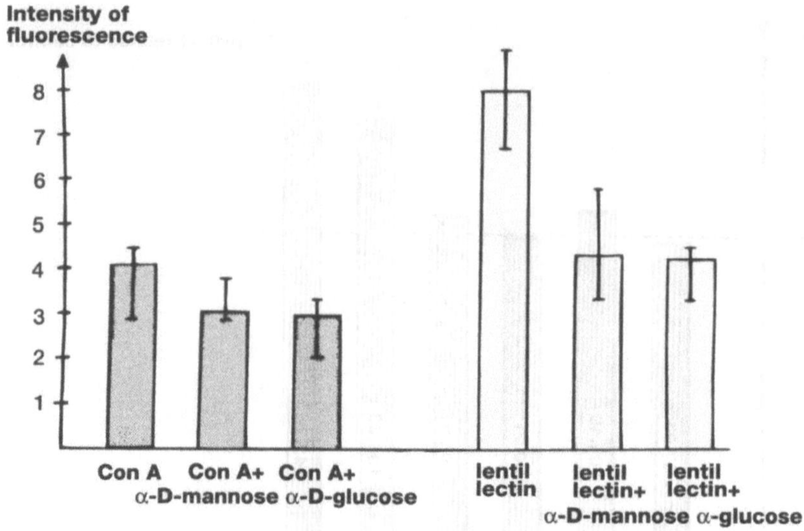

Fig. 12. a-D-glucose and a-D-mannose supersede FITC-labeled Con-A and lentil lectin bound to receptors of *Tetrahymena*. This means that lectins are bound by specific sugars of the cell surface. (Kovács and Csaba 1982)

hormone at the same time or in succession at 24-h intervals, the effect of hormonal imprinting does not decrease for each hormone applied.

5 Sugars of the Receptors

Different lectins are able to bind to *Tetrahymena*. Those abundantly bound are those known to bind simple sugars, while the ones bound to a lesser degree are those with exclusive affinity to amino sugars or sugar oligomeres (Kovács and Csaba 1982). Lectin binding is decreased by giving excessive doses of target sugar (Fig. 12). The most intensively bound lectins are *Pisum sativum* and *Canavalia ensiformis* (Con-A).

Con-A and insulin compete in binding to *Tetrahymena*'s insulin receptor (Kovács et al. 1984; Csaba and Kovács 1982). There is also competition between histamine and Con-A (Csaba and Kovács 1982; Csaba et al. 1983). This means that the sugars of the receptors are also present in *Tetrahymena* and play a role in recognition and binding similar to those in higher-ranked animals.

Insulin imprinting also increases the binding of Con-A. Conversely, imprinting with Con-A does not promote insulin binding. This means that imprinting influences the protein part of the receptor which attracts the appropriate sugars, whose expansion causes enhanced binding of lectin.

6 Cell Aging and Imprinting

The age of the *Tetrahymena* cell culture is a decisive factor for the success of imprinting. Cells formed in the early (logarithmic) phase (18 h) are imprinted well, showing an increased hormone binding after 24 h. When treated with insulin at 42 or 66 h (early and late stationary phase), there is no successful imprinting (Köhidai et al. 1986; Fig. 13). This could mean that aging cells are not able to recognize or fix the presence of the hormone, in other words the membrane is not as dynamic as before.

Tetrahymena cells can be maintained in a special medium under anaerobic condition for months with no need to provide nourishment in a suitable quantity (Williams et al. 1980). This results in the inhibition of cell division and, at the same time, in a multiple prolongation of the original cell's life span.

It was mentioned earlier that under aerobic conditions and in suitable medium *Tetrahymena* remembers the T2 imprinting after 500–600 generations (3 months) or more (Köhidai et al. 1990). Maintained in the anaerobic conditions above, the same situation could be observed, which means that the memory is fresh in the same *Tetrahymena* or in its far progenies alike. Six months after a single treatment under anaerobic conditions, the "memory" is weakened (but the index was significantly higher than in the control), and after 9 months, disappears (Fig. 14). However, cultivated further in normal media, the memory returns after a few days (Csaba et al. 1984d; Fig. 15). This means that the receptor memory developed by the hormonal imprinting is deeply fixed in the cell, possibly requiring a change at gene level.

7 Imprinting by Amino Acids and Oligopeptides

Not only hormones of amino acid and polypeptide type can act at the unicellular level, but also amino acids. Histidine can promote the phagocytosis of *Tetrahymena* as well as histamine (Csaba and Darvas 1986). Tyrosine stimulates cell division as well as T2 (Csaba and Darvas 1987). Pretreatment (im-

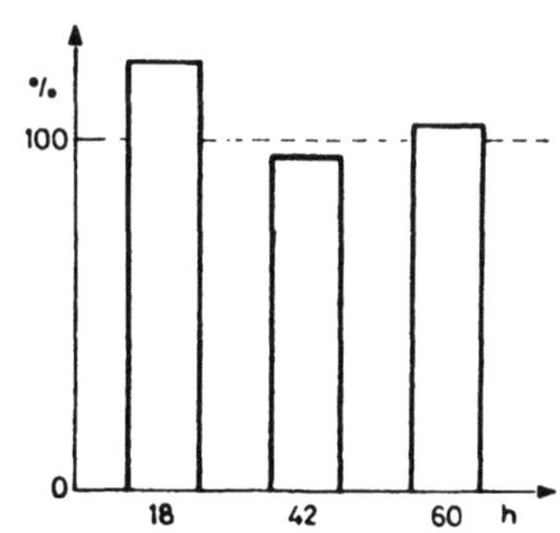

Fig. 13. The imprintability of *Tetrahymena* differs in different periods of cultivation. The highest is in the log phase log phase of reproduction. Control = 100%. (Köhidai et al. 1986)

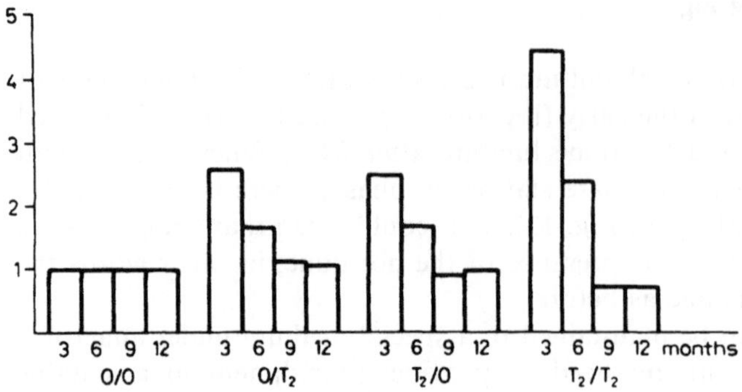

Fig. 14. Growth values of untreated (O); diiodotyrosine (T_2) treated at the end of cultivation period (O/T_2); at the beginning of the cultivation (T_2/O); or twice treated (T_2/T_2) *Tetrahymena* after different periods of cultivation under anaerobic condition. After 3 or 6 months the aged cells react to T_2, their growth is elevated, and after 3 months the growth is more intensive after second T_2 treatment than in the other groups. However, these capacities are lost after 9 months. (Csaba et al. 1984a)

printing) with certain amino acids can increase or decrease the binding of the oligopeptides containing the same amino acids. This makes it likely that the hormone, or at least the amino acid type hormone, receptors of *Tetrahymena* originated from amino acid (food) receptors, as suggested by Lenhoff (1968).

Tetrahymena can select between different oligopeptides. The selection is very sensitive; differences only in one amino acid of six, or a change in the position of an amino acid is enough for the loss or gain of the imprinting effect (Csaba et al. 1986). However, the quality of the amino acid is also important. Proline, an unusual amino acid having the amino group inside the rigid ring structure, seems to have a prominent role in imprinting (Csaba and Kovács 1994a). The associated amino acid is also important: Pro-Phe and Pro-Ala cause strong positive imprinting for themselves, but Pro-Pro and Pro-Leu are negative in this regard (Table 2). Of the seven Tyr-Ala combinations studied, only one provided (negative) imprinting (Csaba and Darvas 1987). These experiments

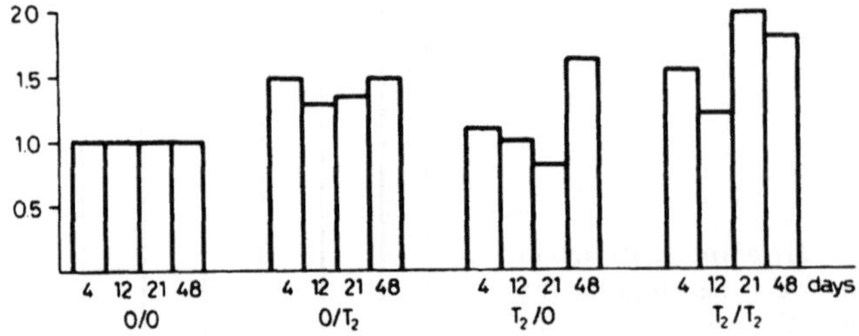

Fig. 15. Growth values of *Tetrahymena* introduced in Fig. 14, transferred to normal medium after 9 months. Abbreviations as in Fig. 14. After 4 days in normal medium, the previously treated cells regain their capacities to grow faster and to react to T_2. (Csaba et al. 1984)

Table 2. Binding of FITC-dipeptides to imprinted *Tetrahymena*

Group number	Imprinted by	Binding of FITC-dipeptide	Control[a] (%)
1	Pro-Ala	Pro-Ala	145.82*
		Pro-Pro	235.20*
		Ala-Ala	54.54*
2	Pro-Leu	Pro-Leu	72.32*
		Pro-Pro	85.35**
		Leu-Leu	64.00*
3	Pro-Phe	Pro-Phe	188.51*
		Pro-Pro	140.46*
4	Pro-Pro	Pro-Pro	85.17**
5	Ser-Ala	Ser-Ala	63.54*
		Ala-Ala	70.41*
6	Ser-Leu	Ser-Leu	75.39*
		Leu-Leu	70.99*
7	Ser-Phe	Ser-Phe	56.02*

[a]Significance: *$p < 0.01$; **$p < 0.5$.

call our attention to the delicate sensitivity of receptor development, which is also supported by the fact that *Tetrahymena* can select between the L and D forms of amino acids (Darvas et al. 1987).

The molecular mass of the peptide does not seem to be very important in provoking imprinting, while the repetition of an amino acid moiety in the sequence of the peptide can influence the imprinting effect. Amino acids and their repetitive sequences not influencing binding or functional characteristics of *Tetrahymena* can imprint it, and in the case of the second encounter there is discrimination between them (in binding and function alike; Csaba and Köhidai 1995).

The presence of positive and negative imprinting points to the differences in receptor binding and cell function. In fact, there is a possibility of durable enhancement in receptor binding by imprinting with amino acids combined with a long-lasting decrease in cell function and vice versa. This fact could be explained by the function of second messenger mechanisms. As a result, the construction of a polypeptide hormone, positive imprint-provoking amino acid composition has been favored.

8 Receptors of the Nuclear Envelope

In mammals, peptide hormone receptors are also present in the nuclear envelope (Horvat 1978; Burwen and Jones 1987; Wong et al. 1988). These receptors can specifically bind the hormones. In *Tetrahymena* the isolated membrane-bound nucleus can also bind insulin. This binding is partly specific and like that of the plasma membrane. The specific binding of labeled hormone could be displaced by an excess of unlabeled one (Csaba and Hegyesi 1992;

Figs. 16, 17). The specificity of the nuclear membrane receptors was also tested by fluorescent labeled antibodies (Csaba and Hegyesi 1994).

After the first encounter (imprinting) with the exogenously given hormone, the receptors' binding capacity increases (Hegyesi and Csaba 1992a,b, 1994). The memory of the first encounter is long-lasting, and increases with time. This means that the nuclear binding sites are imprintable. In intact cells, the binding of insulin to the nuclear membrane is higher than the binding of the plasma membrane. After imprinting, this difference in binding disappears. In addition, 15-min treatment with the hormone is enough to develop maximal value imprinting, which was impossible in the case of the plasma membrane (at least 4 h were needed). This demonstrates not only the stronger binding capacity but also the higher sensitivity of the nuclear membrane to imprinting.

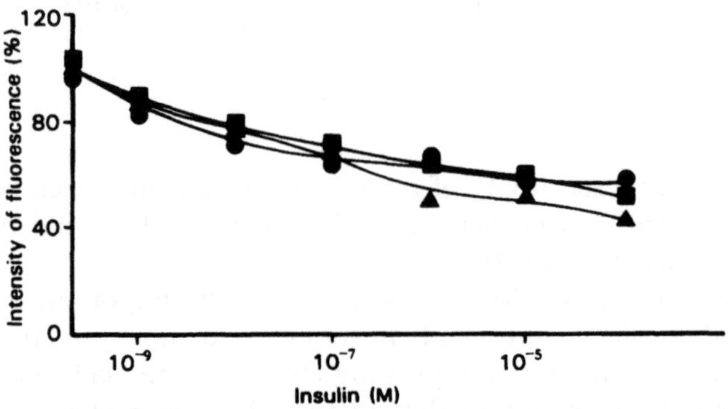

Fig. 16. Binding of FITC-insulin after treatment with different concentrations of unlabeled insulin (displacement) by the nuclei of *Tetrahymena* 24 h (■) and 168 h (▲) after treatament and without treatment (●). There is increasing displacement with progressing time. (Csaba and Hegyesi 1992)

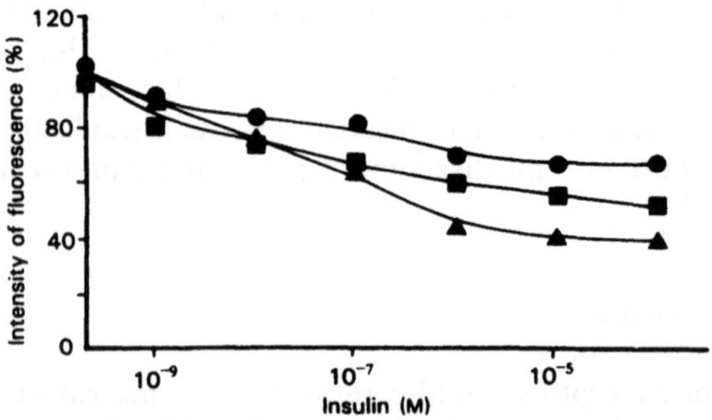

Fig. 17. Binding of FITC-insulin after treatment with different concentrations of unlabeled insulin by the plasma membrane of *Tetrahymena* 24 h (■) and 168 h (▲) after treatment and without treatment (●). The displacement increases parallel with the progressing time. The running of curves is similar to those of Fig. 16, but the differences between the control and imprinted cells are more expressed. (Csaba and Hegyesi 1992)

It is clear that recognition of the environmental elements is extremely important for unicellular organisms, and this justifies the presence of dynamic binding sites in their plasma membrane. However, what accounts for the presence of binding sites in the nuclear envelope? One explanation is the continuous expression and recirculation of the common membrane pool. This could mean that there is a membrane continuity from the nuclear envelope to the plasma membrane, containing the same receptors. However, this can not explain the stronger binding capacity and imprintability of nuclear envelope receptors. To explain this, one has to consider the intracellular insulin production of *Tetrahymena* (Csaba and Hegyesi 1994). The insulin present inside *Tetrahymena* (in the cytoplasm and never in the nucleus) can provoke the development of the nuclear membrane receptors. At the same time, the secreted insulin is immediately so much diluted in the watery milieu that there is no possibility of imprinting by the insulin on its own. Nevertheless, the independence of the two types of receptors is not excluded, taking into account the special membrane pool under the plasma membrane and its turnover.

9 Possible Mechanisms of Imprinting

The exact mechanism of imprinting is unknown. Nevertheless, there are many known factors involved in this mechanism and several hypotheses based upon the known details. When a permanent change at receptorial level, provoked by an effect, is inherited by the progeny generations, different possibilities could be involved in the processes. The first and most likely mechanism would be at a genetic level, that is, the first encounter with the hormone causes a gene rearrangement similar to that produced during the formation of antibody coding genes. However, this mechanism would suppose the genetic effect of a plasma membrane event, and at present, there is no proof available for this mechanism.

The second possible mechanism would be an epigenetic one. Methylation of DNA is responsible for the inactivation of genes. This is also valid for *Tetrahymena,* as insulin binding in 5-azacytidine (a substance inhibiting the maintenance methylase, consequently leading to gene activation)-treated cells increased 24 h after treatment (Csaba and Kovács 1990; Fig. 18). By the rearrangement of methylation pattern of the appropriate genes, the imprinting can influence the expression of receptors, and this change is inherited by the progeny generations.

The third presumable mechanism is connected with the membrane. As was mentioned above, a permanent flow of membrane subpatterns is presumed in the unicellular organisms and the dynamic formation of recognizing membrane patterns in the presence of the signal. Considering a membrane self-assembly after divisions, similar patterns could be formed. However, this does not explain the increased binding of hormones in the progeny.

The role of the membrane in imprinting is supported by the observation that long-lasting starvation diminishes and finally abolishes the possibility of im-

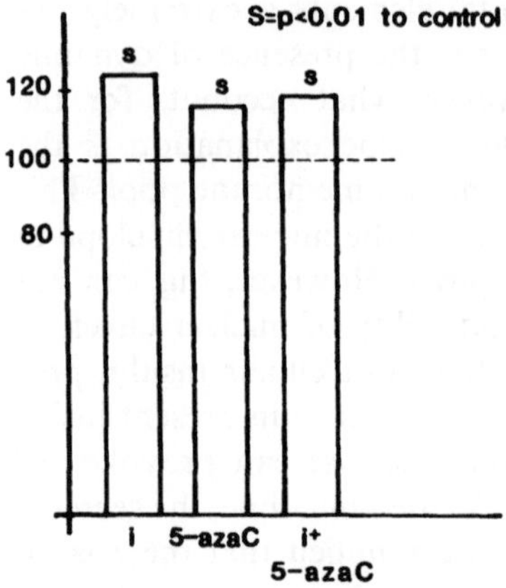

Fig. 18. Effect of insulin, 5-azacytidine, or combined treatment on the insulin binding of *Tetrahymena* 24 h after the treatments. 5-azaC provokes elevation of binding similar to insulin. (Csaba and Kovács 1990)

printing (Csaba et al. 1992; Fig. 19). The starved cell utilizes its nutrient reserves, among them the proteins of membranes. However, there is no irreversible damage to imprintability, and the situation is reversed after permanent cultivation in normal medium. Similarly, the effect of imprinting is lost after cultivating the cells in saline medium, but returns to normal level after cultivating in plain medium. These facts show that the membrane has a very important role in the development of imprinting; however, it does not exclude, but rather reinforces, the necessity for the fixation of the imprinting at a nonmembrane level.

The fourth possibility of transmitting the information of imprinting from the progenitor (imprinted) to progeny cell populations is a "personal" contact

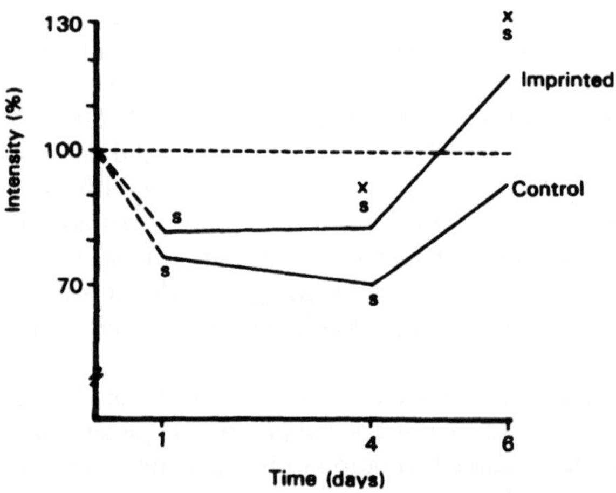

Fig. 19. Insulin binding of *Tetrahymena* cells transferred to saline immediately after imprinting and transferred to normal medium after 1 day. $s = p < 0.01$ to control maintained in normal medium (100%); $x = p < 0.01$ to nonimprinted cells in saline. There is a decrease in binding during starvation and there is no appreciable change for 3 days. However, after that, the untreated cells approach the 100% and the imprinted cells are above it. (Csaba et al. 1992)

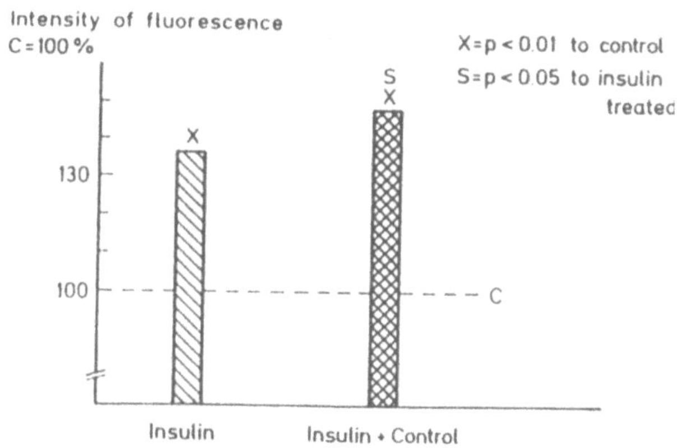

Fig. 20. FITC-insulin binding of insulin-treated and mixed (insulin-treated + control) *Tetrahymena* populations 1 day after treatment. There is significantly higher binding in the mixed cultures. (Csaba and Kovács 1987a)

or the secretion of a transmitter substance (Lawrence et al. 1978; Fletcher et al. 1975; Fletcher and Greenan 1985). If imprinted and intact (virgin) cells are mixed, the hormone-binding capacity of the population is higher than that of the control or imprinted cells (Fig. 20). If the virgin cells cultivated in the filtered (cell-free) medium of imprinted cells, their binding capacity seems similar to the imprinted cells (Figs. 21, 22). Cultivation of virgin cells in the medium of mixed (imprinted and virgin) cells elicits much higher binding capacity (Csaba and Kovács 1987b). This means that there is secretion of some

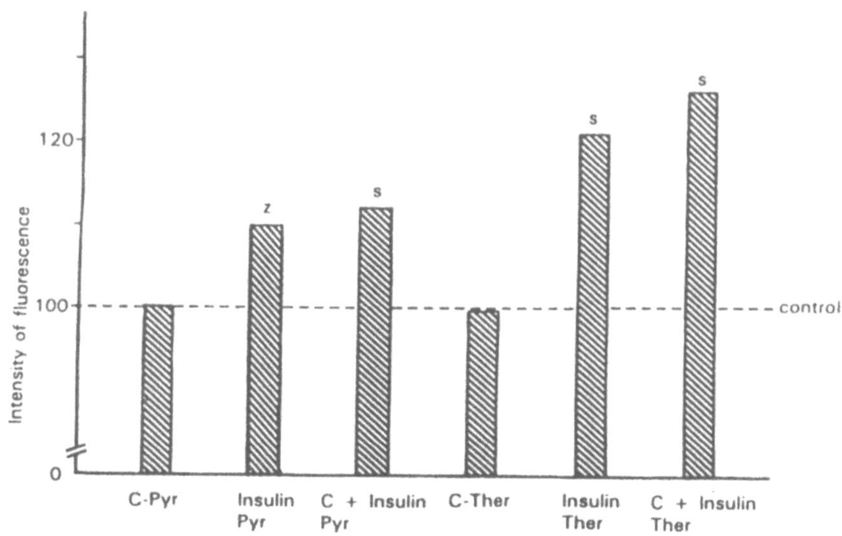

Fig. 21. FITC-insulin binding of (virgin). *T. pyriformis* in the media of the different (pretreated) groups, related to the control-medium-treated groups (*C-Pyr*; *C-Ther*) as 100. The media of the insulin-pretreated and mixed (*C + insulin*) groups increased the hormone binding of virgin cells, irrespective of the donor taxon. $s = p < 0.01$; $z = p < 0.05$. (Csaba and Kovács 1987a)

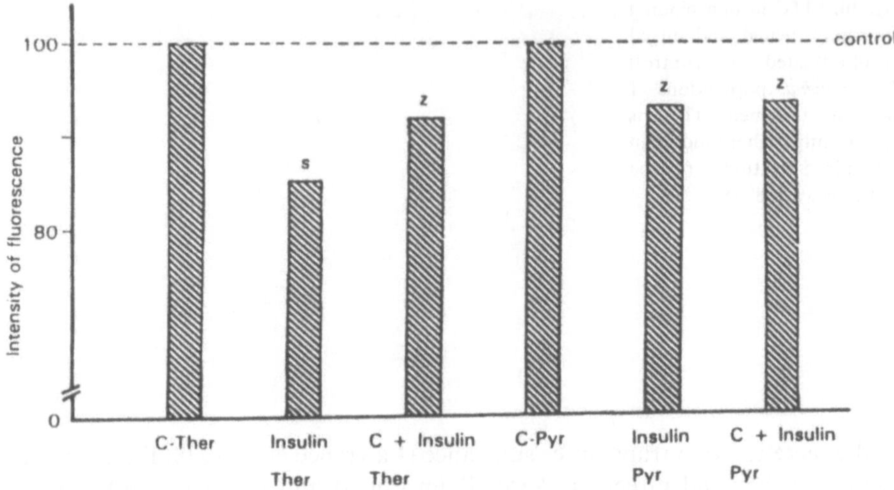

Fig. 22. FITC-insulin binding of (virgin) *T. thermophila* in the media of different (pretreated) groups, related to the control-medium-treated groups (*C-Pyr*; *C-Ther*) as 100. The media of the insulin-pretreated and mixed (*C+insulin*) groups decreased the hormone binding of virgin cells, irrespective of the donor taxon. $s = p < 0.01$; $z = p < 0.05$. (Csaba and Kovács 1987a)

imprinting transmitter substances, the quantity of which increases in the presence of virgin cells. There is thus a possibility of the transmission of receptor memory by secretion; however, it is not certain that this mechanism can function under natural conditions, where the concentration of substances secreted is extremely low.

If we accept that transmission of imprinting is dependent on the transmitting substances, then the importance of postreceptorial events much be emphasized. In *Tetrahymena* (T.) *pyriformis*, insulin imprinting provokes a

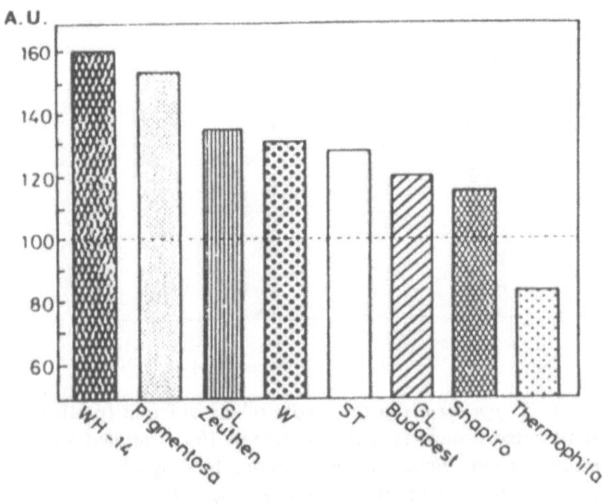

Fig. 23. Binding of FITC-labeled insulin to *Tetrahymena* pretreated (imprinted) with insulin. Values related to the not-pretreated control as 100%. All the taxa beneath the columns reacted with positive imprinting, except *T. thermophila* which displayed negative imprinting; *A.U.* arbitrary units (Kovács et al. 1982)

positive imprinting (as in many other taxa); however, as an exception, negative imprinting is provoked in *Tetrahymena thermophila* (Csaba and Kovács 1987a; Fig. 23). If *T. thermophila* is cultivated in the medium of imprinted *T. pyriformis*, negative imprinting is provoked, and vice versa, culturing of *T. pyriformis* in the medium of imprinted *T. thermophila* results in positive imprinting (Figs. 21, 22). This implies that the trend of imprinting is determined by intracellular events.

The receptorial and intracellular (postreceptorial) events are important not only in the inheritance of but also in fulfilling imprinting. Any disturbance of the membrane fluidity (for instance by ethanol, local anesthetics, etc.; Nozawa et al. 1985a,b) or inhibition of second messenger systems (Kovács and Csaba 1992b; Kovács et al. 1988) diminishes or completely inhibits the development of imprinting. However, these processes do not interfere with the transmission of receptor memory to the progeny generations.

10 The Other Component: Hormones in Protozoa

It was not surprising to find hormone receptors in unicellular organisms, considering the fact that the ability of recognition plays an important role in their survival. However, it was a surprise to observe that these organisms also contain hormones (hormone-like materials) characteristic of higher vertebrates. Using immunological and pharmacological methods, the presence of polypeptide hormones, such as insulin, somatostatin, adrenocorticotropic hormone, relaxin, arginine vasotocin, vasopressin, calcitonin, endothelin, and beta endorphine was demonstrated in *Tetrahymena* (LeRoith et al. 1980, 1981, 1982, 1983; Roth et al. 1982; Lenard 1992; Köhidai and Csaba 1995). Most of these hormones were not only present in the cells, but were secreted by them. Amino acid-type hormones such as serotonin (Csaba and Kovács 1994a) and epinephrine (Blum 1976; Brizzi and Blum 1970; Kariya et al. 1974: Goldman et al. 1981) were also found in *Tetrahymena*. The presence of steroid hormones, such as estradiol and dihydroepiandrosterone, was also pointed out (Csaba et al. 1985a). The presence of prostaglandins (Csaba and Nagy 1987) was dubious, and triiodothyronine, thyroxine (Csaba and Nagy 1987), and atrial natriuretic peptide (Köhidai et al. 1995) were not proved; however, they were checked.

In other unicellular organisms, such as fungi (*Neurospora crassa*, *Candida albicans*, *etc.*), the presence of insulin or steroid hormones was also demonstrated (Fawell et al. 1988; LeRoith et al. 1987; Lenard 1992). Some of the bacteria studied also contain hormone-like materials (LeRoith et al. 1981, 1987).

Hormones are messenger molecules which transmit information between the cells of the same organism. However, in unicellular organisms, the cell and the organism are identical. This means that transmission of information would be intracrine or autocrine, or could aim at another unicellular organism. Considering the quantity of water in the surrounding milieu, intercellular transmission of information does not seem to be likely under natural conditions. However, intracriny of autocriny cannot be disregarded, considering the presence of receptors in the plasma membrane and nuclear envelope.

Why are many different kinds of hormones present in *Tetrahymena*? They could be no more than trials of evolution, that is, the amino acid or peptide-type hormones are side products of a broad-spectrum protein synthesis of a highly developed ciliate organism. From this point of view, there is a possibility of hormone selection for Metazoa as well as the selection of receptors.

11 Evolutionary Conclusions Based on the Unicellular Model

The basis for a living biological recognition system is the relation between the cell and its environment. This environment would also be living since other cells or non-living molecules produced by cells or nature are also present. Therefore, the endocrine systems of vertebrates should be considered, from an evolutionary viewpoint, as recognition systems where the signal, the channel containing and transporting the signal, and the signal receiver are of decisive importance.

If we consider *Tetrahymena* as a cell, and not as an organism, then it contains the receiver, the channel is the water around it, and myriads of molecules are circulating in the water as signals. To *Tetrahymena* the signals mean "nourishment", "danger", "attraction", or "repulsion", and the cell has to recognize them if it is to remain alive. After the reception of the signals, the cell should react; this requires the function of second messenger systems. As a result, signal recognition and signal transmitting systems are universal systems of absolute importance in the life of an animal cell.

The sensitivity of the receivers must be high enough to avoid higher concentrations of toxic substances. As was mentioned earlier, the insensitive cells are not able to transmit this characteristic to the progeny. The high sensitivity of receptors developed at this low level of phylogeny is manifested in vertebrate receptors.

It has to be emphasized that the protozoan cells do not recognize (vertebrate) hormones, but rather recognize molecules. These molecules could be well or less recognizable, and could provoke insignificant or important physiological reactions, or none at all. After the development of receptor-like structures, the cell could differentiate the molecules according to its requirements. If a (signal) molecule is able to promote such processes as are advantageous to the protozoan cell, a signal-receiver connection develops which is equal to a hormone-receptor connection (Csaba 1980, 1986a; Figs. 24, 25, 26). This selection and this connection also could be manifested in the metazoan organism. In these organisms the receptors are specific targets of specific signal molecules; however, by chance, other molecules can also bind them, e.g., viruses, lectins, etc.

Fig. 24. Influence of hormones and a second messenger on the interval between two systoles of the contractile vacuole of Tetrahymena related to the control (100%). *The higher the column,* the longer the time. Oxytocin increased the interval extremely, while vasopressin was ineffective. (Csaba and Kovács 1992)

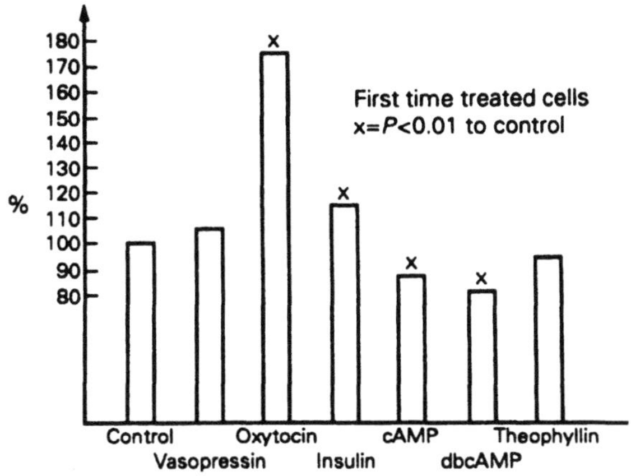

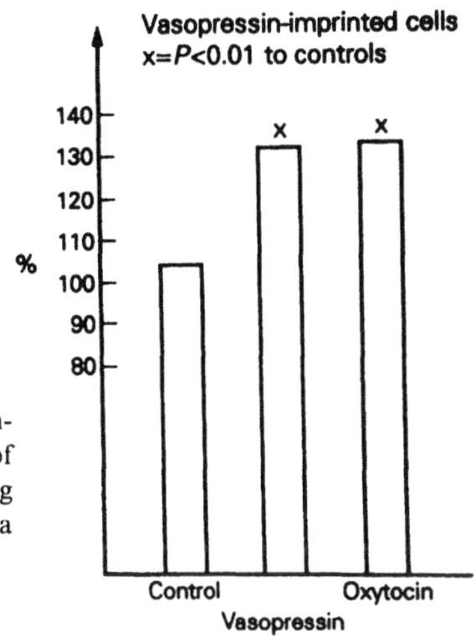

Fig. 25. Influence of vasopressin and oxytocin on the interval between two systoles of the contractile vacuole of *Tetrahymena* after vasopressin imprinting. The imprinting equalized the effect of the two related hormones. (Csaba and Kovács 1992)

The proper structure (sequence and steric configuration) of a molecule is the prerequisite for it to be selected as a signal (hormone). In vertebrate organisms, a few polypeptide hormone families dominate regulation and only a few amino acids developed as hormones, and cholesterol is the basis of most of the morphogenetic hormones. This makes it likely that certain amino acids and amino acid sequences (or lipids) were more suitable structures for signal molecules (and they were not occupied in other functions) than others. This is unequivocally demonstrated also at a unicellular level. Amino acids and amino acid compositions influence the functions of *Tetrahymena* differently, and they are also different in their imprinting capacity. Possibly, this latter was responsible for the selection of the molecules most suitable to function as hormones. However, at a primitive level only the outlines of the hormones were shaped, an evolutionary period was needed to work out the definitive vertebrate form. This is justified by the enhanced sensitivity of *Tetrahymena* and

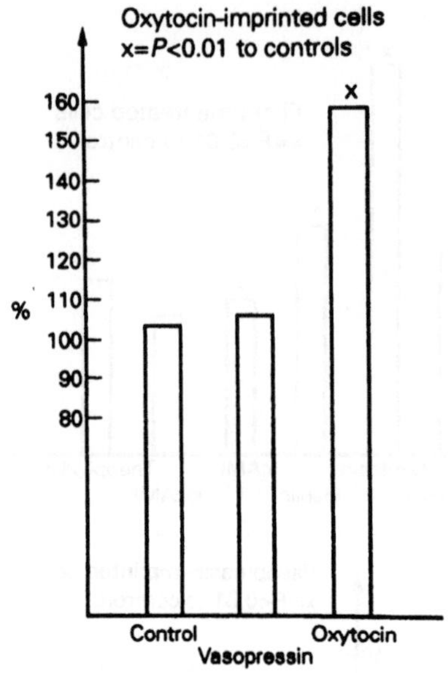

Fig. 26. Influence of vasopression and oxytocin on the interval between two systoles of the contractile vacuole of *Tetrahymena* after oxytocin imprinting. The imprinting favorized oxytocin without influencing the effect of vasopressin. (Csaba and Kovács 1992)

invertebrates (e.g., planarians: Csaba and Kádár 1978) to the precursors of vertebrate hormones (Csaba et al. 1980). From this point of view, the receptor seems to be more conservative and the hormone develops faster. The more primitive receptor can recognize the "most developed" appropriate hormone, with hardly any physiological effect at this level (Muggeo et al. 1979a,b). The overlap (cross-reaction) between hormones on a receptor is not more expressed in *Tetrahymena* than in mammals.

Chemical imprinting is a tool to enhance sensitivity. The cell becomes more sensitive after imprinting, and this enhancement of sensitivity is manifested in the increased binding capacity of receptors and in the change in behavior of second messenger systems. Imprinting is a possibility for enhancement of recognition at a unicellular level and a necessity for receptor function at the metazoan level (Csaba 1984, 1991, 1994). This means that chemical imprinting is a universal phenomenon at different levels of evolution.

At a unicellular level, continuous monitoring of the immediate environment is needed, as a consequence of the continuously changing milieu. At a metazoan level, the receptorial recognition serves the internal milieu, which is relatively constant, and signal molecules are determined. For this purpose, determined receptors are needed as well. At the unicellular level, the abundance of permanently changing receivers of the same cell serves the exactness of recognition; at a metazoan level, the predetermined different receptors of different cells are responsible for the homeostasis of the organism. Many receptors in one cell, representing all the functions, diverged to a few receptors in many cells during different functions. Nevertheless, in mammals, for example, during perinatal maturation, the predetermined receptors require enforcement by hormonal imprinting, as is required by *Tetrahymena*'s binding structures (Csaba 1984, 1986b, 1991, 1994).

If we accept Koch's (1979) blind automation theory; that is, the cell at a low level of phylogeny randomly produces recognition structures. and we know that an enormous variation of molecules can be present in the milieu, the encounter of the to-be receptor and the to-be hormone is a chance event. This fortuity, representing a completely open system, develops into a determination in metazoa, when the information (hormone)-producing cell and the receiver (receptor)-containing cell are settled in the same organism, which also has predetermined channels. The informator and target cells have the same genome; however, the information is disparately manifested.

Chemical imprinting is regularly transmitted to the progeny generations. This is well understandable in *Tetrahymena*, where the duration of individual life is very short, and the transmission of imprinting works as a sort of "cultural evolution". However, in metazoa, transmission is also needed considering the perinatal time of imprinting and the limited life span – longer, however, than that of the *Tetrahymena* – of the cells directly imprinted. This means that the transmission of imprinting has been fixed at a unicellular level and has been maintained during evolution.

Considering these facts and hypotheses, it can be concluded that the basic elements of receptor-hormone recognition are present at the lowest level of phylogeny, in protozoa, and at a very primitive level of metazoan life (Gamulin et al. 1994; Müller et al.; Müller 1995). These elements are employed for further evolution of this system, based on the necessities of the cells and differential distribution in cells. Accordingly, as phagocytosis and ciliary movement are traceable at a unicellular level, in the same manner, cellular recognition and its associated processes can also be observed at this level.

Acknowledgments. The work reviewed in this chapter was supported by Hungarian National Research Fund (T-013355) and the Hungarian Ministry of Welfare. The author is indebted to Elsevier Sciences Ltd, Academic Press, Faculty Press, Verlag fur Zeitschrift fur Naturforschung, Cell Biochemistry and Function, Acta Protozoologica, S. Karger AG, Birkhäuser Verlag, Bioscience Reports, and the Publishing House of Hungarian Academy of Sciences for their kind permission to reuse material published earlier in their journals.

References

Azukizawa M, Kutzman G, Pekary AE, Hershman JK (1977) Comparison of the binding characteristics of bovine thyrotropin and human chorionic gonadotropin to thyroid plasma membrane. Endocrinology 101: 1880–1889

Blum JJ (1967) An adrenergic control system in *Tetrahymena*. Proc Natl Acad Sci USA 51: 81–88

Brizzi G, Blum JJ (1970) Effect of growth conditions on serotonin content of *Tetrahymena pyriformis*. J Protozool 17: 553–555

Burwen SJ, Jones AL (1987) The association of polypeptide hormones and growth factor with the nuclei of target cells. Trends Biochem Sci 12: 159–162

Christopher GK, Sundermann CH (1992) Conventional and confocal microscopic studies of insulin receptor induction in *Tetrahymena pyriformis*. Exp Cell Res 201: 477–484

Csaba G (1980) Phylogeny and ontogeny of hormone receptors: the selection theory of receptor formation and hormonal imprinting. Biol Rev 55: 47–63

Csaba G (1981) Ontogeny and phylogeny of hormone receptors. Karger, Basel

Csaba G (1984) The present state in the phylogeny and ontogeny of hormone receptors. Horm Metab Res 16: 329–335

Csaba G (1985) The unicellular *Tetrahymena* as a model cell for receptor research. Int Rev Cytol 95: 327–377

Csaba G (1986a) Development of hormone receptors. Why do hormone receptors arise? Experientia 42: 715–719

Csaba G (1986b) Receptor ontogeny and hormonal imprinting. Experientia 42: 750–759

Csaba G (1991) Interaction between the genetic programme and environmental influences in the perinatal critical period. Zool Sci 8: 813–825

Csaba G (1994) Phylogeny and ontogeny of chemical signalling: origin and development of hormone receptors. Int Rev Cytol 155: 1–48

Csaba G, Darvas Zs (1984) Influence of plant hormones (indoleacetic acid, gibberelline, naphtylacetic acid and kinetine) on the growth of *Tetrahymena pyriformis*. Cell Biol Int Rep 8: 181

Csaba G, Darvas Zs (1986) Receptor-level interrelationships of amino acids and the adequate amino acid type hormones in *Tetrahymena*: a receptor evolution model. Biosystems 19: 55–59

Csaba G, Darvas Zs (1987) Hormone evolution studies: multiplication, promotion and imprinting ("memory") effects of various amino acids on *Tetrahymena*. BioSystems 20: 225–229

Csaba G, Hegyesi H (1992) Specificity of insulin binding by plasma and nuclear membrane receptors in *Tetrahymena*. Similarities and dissimilarities at the two levels. Cytobios 70: 153–158

Csaba G, Hegyesi H (1994) Immunocytochemical verification of the insulin receptor's specificity in the nuclear envelope of *Tetrahymena*. Comparison with receptor of the plasma membrane. Biosci Rep 14: 25–31

Csaba G, Kádár M (1978) The possibility of the amplification of hormone receptors in early periods of differentiation. The invertebrate model. Differentiation 10: 61–64

Csaba G, Kovács P (1982) Histamine-lectin and insulin-lectin binding site overlap in *Tetrahymena*. Cell Mol Biol 28: 153–158

Csaba G, Kovács P (1987a) Taxon dependence of receptor level cell-to-cell communication in *Tetrahymena*: possible explanation for the transmission of hormonal imprinting. Cytobios 52: 17–22

Csaba G, Kovács P (1987b) Transmission of hormonal imprinting in *Tetrahymena* cultures by intercellular communication. Z Naturforsch 42C: 932–934

Csaba G, Kovács P (1990) Impact of 5-azacytidine on insulin binding and insulin-induced receptor formation in *Tetrahymena*. Biochem Biophys Res Comun 168: 709–713

Csaba G, Kovács P (1991) EGF receptor induction and insulin-EGF overlap in *Tetrahymena*. Experientia 47: 718–721

Csaba G, Kovács P (1992) Oxytocin and vasopressin change the activity of the contractile vacuole in *Tetrahymena*: new contributions to the phylogeny of hormones and hormone receptors. Comp Biochem Physiol 102A: 353–355

Csaba G, Kovács P (1994a) Effect of hormones and hormone-induced imprinting on the serotonin level in *Tetrahymena*: immunocytochemical studies. Microbios 80: 155–163

Csaba G, Kovács P (1994b) Role of proline in the imprinting developed by dipeptides in *Tetrahymena*. Possible role in hormone evolution. Experientia 50: 107–109

Csaba G, Köhidai L (1986) Modeling the insulin receptor in the *Tetrahymena*. Time dependence of receptor function, down regulation and imprinting. Acta Protozool 25: 331–338

Csaba G, Köhidai L (1995) Effects of alanin and alanine oligopeptides on the chemotaxis of *Tetrahymena*. Evolutionary conclusions. Biosci Rep 15: 185–190

Csaba G, Lantos T (1973) Effect of hormones on protozoa. Studies on the phagocytotic effect of histamine, 5-hydroxytriptamine and indoleacetic acid on *Tetrahymena pyriformis*. Cytobiologie 7: 361–365

Csaba G, Lantos T (1975) Effect of insulin on the glucose uptake of protozoa. Experientia 31: 107–109

Csaba G, Lantos T (1976) Effect of epinephrine on glucose metabolism in *Tetrahymena*. Endokrinologie 68: 239–240

Csaba G, Nagy SU (1976) Effect of vertebrate hormones on the cyclic AMP level in *Tetrahymena*. Acta Biol Med Germ 35: 1399–1401

Csaba G, Nagy SU (1987) Presence (HPL, prostaglandin) and absence (triiodothyronine, tyroxine) of hormones in *Tetrahymena*: experimental facts and open questions. Acta Physiol Hung 70: 105–110

Csaba G, Németh G (1980) Effect of hormones and their precursors on protozoa – the selective responsiveness of *Tetrahymena*. Comp Biochem Physiol 65B: 387–390

Csaba G, Ubornyák L (1982a) Effect of polypeptide hormones (insulin, thyrotropin, gonadotropin, adrenocorticotropin) on RNA synthesis in *Tetrahymena*, as assessed from incorporation of 3H-uridine. Acta Biol Hung 33: 381–384

Csaba G, Ubornyák L (1982b) Effect of plant hormones (IAA, NAA and TAA) on the 3H-uridine incorporation of *Tetrahymena*. Acta Biol Hung 33: 423–424

Csaba G, Nagy SU, Lantos T (1976) Are biogenic amines acting on *Tetrahymena* through a cyclic AMP mechanism? Acta Biol Med Germ 35: 259–261

Csaba G, Bierbauer J, Fehér S (1980) Influences of melatonin and its precursors on the pigment cells of Planaria (*Dugesia lugubris*). Comp Biochem Physiol 67C: 207–209

Csaba G, Németh G, Vargha P (1982) Influence of hormone concentration and time factor on development of receptor memory in a unicellular (*Tetrahymena*) model system. Comp Biochem Physiol 73B: 357–360

Csaba G, Darvas Zs, László V (1983a) A functional study of concanavalin A-histamine binding site overlap in *Tetrahymena phagocytosis* test. Comp Biochem Physiol 75A: 457–460

Csaba G, Németh G, Vargha P (1983b) Attempt to disturb receptor memory in a unicellular (*Tetrahymena*) model system. Acta Physiol Hung 61: 131–136

Csaba G, Darvas Zs, László V, Vargha P (1984a) Influence of prolonged life span on receptor memory in a unicellular organism, *Tetrahymena*. Exp Cell Biol 52: 211–216

Csaba G, Kovács P, Inczefi-Gonda Á (1984b) Insulin binding sites induced in the *Tetrahymena* by rat liver receptor antibody. Z Naturforsch 39C: 183–185

Csaba G, Muzsnai Á, László V, Darvas Zs (1984c) A new model for the quantitative determination of phagocytosis in amoeba. Effect of histamine on the phagocytotic activity of *Amoeba proteus*. Cytologia 49: 691–695

Csaba G, Németh P, Vargha P (1984d) Receptor memory in *Tetrahymena*; does it satisfy the general criteria of memory? Expl Cell Biol 52: 320–325

Csaba G, Inczefi-Gonda Á, Fehér T (1985a) Induction of steroid binding sites (receptors) and presence of steroid hormones in the unicellular *Tetrahymena pyriformis*. Comp Biochem Physiol 82A: 567–570

Csaba G, Németh G, Kovács P, Vargha P, Vas Á (1985b) Chemical reception mechanism at a low level of phylogeny. Influence of polypeptide hormones and non-hormone polypeptides on the growth of *Tetrahymena*. BioSystems 17: 227–231

Csaba G, Kovács P, Török O, Bohdaneczky E, Bajusz S (1986) Suitability of oliogopeptides for induction of hormonal imprinting – implications on receptor and hormone evolution. Biosystems 19: 285–288

Csaba G. Kovács P, Vas Á (1987) Effect of cAMP level on hormonal imprinting in *Tetrahymena*. Acta Physiol Hung 69: 231–237

Csaba G. Fülöp AK, Inczefi-Gonda Á (1989) Presence of benodiazepine binding sites (receptors) and amplification thereof by imprinting in *Tetrahymena*. Experientia 45: 96–98

Csaba G, Kovács P, Klein I (1992) Impact of starvation on hormone binding and hormonal imprinting in *Tetrahymena*. Cytobios 69: 7–13

Csaba G, Kovács P, Falus A (1995) Human cytokines interleukin (IL)-3 and IL-6 affect the growth and insulin binding of the unicellular organism *Tetrahymena*. Cytokine (7: 771–774)

Darvas ZS, Nozawa Y, Csaba G (1987) Dissimilar effect of L and D amino acids on the growth of *Tetrahymena*. Biosci Rep 7: 757–760

Fawell SE, McKenzie MA, Greenfield NJ, Adebodun F, Jordan F, Lenard J (1988) Stimulation by mammalian insulin of glycogen metabolism in a wall-less strain of *Neurospora crassa*. Endocrinology 122: 518–523

Fletcher WH, Greenan JRT (1985) Receptor mediated action without receptor ocupancy. Endocrinology 116: 1660–1662

Fletcher WH, Anderson NC, Everett JW (1975) Intercellular communication in the rat anterior pituitary gland. J Cell Biol 67: 467–476

Gamulin V, Rinkevich B, Schäcke H, Kruse M, Müller IM, Müller WEG (1994) Cell adhesion receptors are highly conserved from the lowest metazoa (marine sponges) to vertebrates. Biol Chem Hoppe-Seyler 375: 583–588

Ginsberg BH, Kahn CR, Roth J (1977) The insulin receptor of the turkey erythrocyte, similarity to mammalian insulin receptor. Endocrinology 100: 520–525

Goldman ME, Gundersen R, Erickson CK, Thompson GA (1981) High performance liquid chromatography analysis of catecholamines in growing and non-growing *Tetrahymena*. Biochem Biophys Acta 676: 221–225

Gorbman A, Dickhoff WW, Vigna SR, Clark NB, Ralph CL (1983) Comparative endocrinology. Wiley, New York

Hegyesi H, Csaba G (1992a) Specific insulin binding by and imprintability of the nuclear membrane of *Tetrahymena*. Cytobios 72: 153–157

Hegyesi H, Csaba G (1992b) Effect of pretreatment (imprinting) with different concentrations of insulin binding of nuclear and plasma membrane in *Tetrahymena*. Cytobios 72: 191–196

Hegyesi H, Csaba G (1994) Changes in insulin binding capacity of plasma membrane and nuclear envelope of *Tetrahymena* following a single long-term pretreatment and several short-term pretreatments (imprinting) with insulin. Cytobios 79: 189–194

Horvat A (1978) Insulin binding sites on rat liver nuclear membranes: biochemical and immunofluorescence studies. J Cell Physiol 97: 37–40

Kariya K, Saito K, Iwata H (1974) Adrenergic mechanism in *Tetrahymena*. cAMP and cell proliferation. Jpn J Pharmacol 24: 129–134

Koch AS, Fehér J, Lukovics I (1979) Single model of dynamic receptor pattern generation. Biol Cybern 32: 125–138

Kovács P, Csaba G (1982) Studies on the lectin binding capacity of *Tetrahymena*. Acta Protozool 21: 69–75

Kovács P, Csaba G (1985) Receptor level study of polypeptide hormone (insulin, TSH, FSH) imprinting an overlap in *Tetrahymena*. Acta Protozool 24: 37–40

Kovács P, Csaba G (1987) The role of Ca^{2+} in hormonal imprinting of the *Tetrahymena*. Acta Physiol Hung 69: 167–169

Kovács P, Csaba G (1990a) Influence of the phosphonositol (PI) system in the mechanism of horomonal imprinting. Biochem Biophys Res Commun 170: 119–126

Kovács P, Csaba G (1990b) Effect of insulin on *Blepharisma undulans* (Stein) at primary exposure and reexposure. Acta Protozool 29: 131–139

Kovács P, Csaba G (1990c) Evidence of the receptor nature of the binding sites induced in *Tetrahymena* by insulin treatment. A quantitative cytofluorimetric technique for the study of binding kinetics. Cell Biochem Funct 8: 49–56

Kovács P, Csaba G (1992a) Effect of inhibition and activation of tyrosin kinase on insulin imprinting in *Tetrahymena*. Cell Biochem Funct 10: 267–271

Kovács P, Csaba G (1992b) Effect of tyrosin kinase inhibitor tyrphostin on insulin binding and insulin imprinting of *Tetrahyhmena*. Microbios 79: 37–41

Kovács P, Csaba G, László V (1984) Study on the imprinting and overlap of insulin and Concanavalin A at the receptor level in protozoan (*Tetrahymena*) model system. Acta Physiol Hung 64: 19–23

Kovács P, Csaba G, Bohdaneczky E (1985) Immunological evidence of the induced insulin receptor in *Tetrahymena*. Comp Biochem Physiol 80A: 41–42

Kovács P, Nozawa Y, Csaba G (1988) Effect of neomycin on hormonal imprinting in *Tetrahymena*. Microbios Lett 39: 67–70

Kovács P, Nozawa Y, Csaba G (1989) Induction of hormone receptor formation in the unicellular *Tetrahymena*. Bioxi Rep 9: 87–92

Köhidai L, Csaba G (1995) Effect of mammalian vasoconstrictor peptide endothelin-1 on the unicellular *Tetrahymena*. Immunocytochemical presentation of endogenous endothelin activity. Comp Biochem Physiol 111C: 311–316

Köhidai L, Thomka M, Csaba G (1986) Age of the cell culture: a factor influencing hormonal imprinting of *Tetrahymena*. Acta Microbiol Hung 33: 295–300

Köhidai L, Csaba G, László V (1990) Persistence of receptor "memory" induced in *Tetrahymena* by insulin imprinting. Acta Microbiol Hung 37: 269–275

Köhidai L, Kihara A, Csaba G (1992) Correlation of insulin pretreatment and insulin binding of *Amoeba proteus*: a new technique for evaluation. Comp Biochem Physiol 103A: 535–539

Köhidai L, Csaba G, Karsa J (1995) Effect of atrial natriuretic peptide (ANP) on the unicellular model *Tetrahymena pyriformis*. Microbios 82: 27–40

Kuno T, Yoshida N, Tanaka C (1979) Immunocytochemical localization of cyclic AMP and cyclic GMP in synchronously dividing *Tetrahymena*. Acta Histochem Cytochem 12: 563

Lawrence TS, Beers W, Gilula NB (1978) Transmission of hormonal stimulation by cell-to-cell communication. Nature 272: 501–506

Legros F, Uidenhoef P, Dumont J, Hanson B, Jeanmart J, Massant B, Conard V (1975) Specific binding of insulin to the unicellular alga *Acetabularia mediterranea*. Protoplasma 86: 119–137

Lenard J (1992) Mammalian hormones in microbial cells. Trends Biochem Sci 17: 147–150

Lenhoff HM (1968) Behaviour, hormones and hydra. Science 161: 434–442

LeRoith D, Schiloach J, Roth J, Lesniak MA (1980) Evolutionary origins of vertebrate hormones: substances similar to mammalian insulin are native to unicellular eukaryotes. Proc Natl Acad Sci USA 77: 6184–6186

LeRoith D, Schiloach J, Roth J, Lesniak MA (1981) Evidence that material very similar to insulin is native to E.coli. J Biol Chem 256: 6533–6536

LeRoith D, Liotta AH, Roth J, Schiloach J, Lewis ME, Pert CB, Krieger DT (1982) ACTH and beta endorphin like materials are native to unicellular organism. Proc Natl Acad Sci USA 79: 2086–2090

LeRoith D, Shiloach J, Berelowitz M, Frohmann LA, Liotta AS, Krieger BT, Roth J (1983) Are messenger molecules in microbes the ancestors of the vertebrate hormones and tissue factor? Fed Proc 42: 2602–2607

LeRoith D, Roberts C Jr, Lesniak MA, Roth J (1987) Receptor for intercellular messenger molecules in microbes: similarity to vertebrate receptors and possible implications for diseases in man. In: Csaba G (ed) Development of hormone-receptors. Birkhäuser, Basel, 167 pp

McKenzie MA, Fawell SE, Cha M, Lenard J (1988) Effect of mammalian insulin on metabolism growth and morphology of a wall-less strain of *Neurospora crassa*. Endocrinology 122: 511–517

Muggeo M, Ginsberg GH, Roth J, Neville GH, de Meyts P, Kahn CR (1979a) The insulin receptor in vertebrates is functionally more conserved during evolution than the insulin itself. Endocrinology 104: 1393–1402

Muggeo M, van Obberghen E, Kahn CR, Roth J, Ginsberg BA, de Meyts P, Emdin SO, Falkmer S (1979b) The insulin receptor and insulin of the atlantic hagfish. Diabetes 28: 175–181

Müller WEG (1995) Molecular phylogeny of Metazoa (animals): monophyletic origin. Naturwissenschaften 82: 321–329

Müller WEG, Müller IM, Rinkeich B, Gamulin V (1995) Molecular evolution: evidence for the monophyletic origin of multicellular animals. Naturwissenschaften 82: 36–38

Nozawa Y, Kovács P, Csaba G (1985a) Influence of membrane perturbation elicited after hormone treatment (hormonal imprinting) on the later hormone binding capacity of *Tetrahymena pyriformis*. Cell Mol Biol 31: 13–16

Nozawa Y, Kovács P, Csaba G (1985b) The effects of membrane perturbants, local anesthetics and phenothiasines on hormonal imprinting in *Tetrahymena pyriformis*. Cell Mol Biol 31: 223–227

O'Neil JB, Pert CB, Ruff MR, Smith CC, Higgins WJ, Zipser B (1988) Identification and characterization of the opiate receptor in the ciliated protozoan *Tetrahymena*. Brain Res 450: 303–315

Roth J, Le Roith D, Shiloach J, Rosenzweig AL, Lesniak MA, Havrankova J (1982) The evolutionary origins of hormones, neurotransmitters and the extracellular messengers. N Engl J Med 306: 523–527

Williams NE, Wolfe J, Bleyman LK (1980) Long-term maintenance of *Tetrahymena*. J Protozool 27: 327–328

Wong KY, Hawley D, Vigneri R, Goldfine ID (1988) Comparison of solubilised and purified plasma membrane and nuclear insulin receptors. Biochemistry 27: 375–379

Zipser B, Ruff MR, O' Neil JB, Smith CC, Higgins WJ, Pert CB (1988) The opiate receptor: A single 111 kDa recognition molecule appears to the conserved in *Tetrahymena*, leech and rat. Brain Res 463: 296–304

Studies on the Opioid Mechanism in *Tetrahymena*

F.L. Renaud[1], R. Chiesa[1], F. Rodríguez[1], N. Tomassini[1], and M. Marino[2]

1 Introduction to Opioid Mechanisms

The group of compounds collectively called opioids includes both opiate alkaloids of medical importance such as morphine and heroin and endogenous opioid peptides such as met- and leu-enkephalin, β-endorphin, and dynorphin (Simon and Hiller 1994). Opioid mechanisms have been studied primarily in the mammalian nervous system, although there is also evidence for their presence in cells of the immune system (Fischer 1988). In general, the physiological effect of these compounds is inhibitory, and this inhibition is exerted by opioid receptors that have been classified as μ, δ, and κ according to their affinity for specific agonists (Simon and Hiller 1994). These receptors transduce the opioid signal by means of a G-proteins which then modulate an effector molecule, such as adenylate cyclase and ion channels (North et al. 1987; Childers 1991; Laugwitz et al. 1993). We shall now review briefly how this situation compares with what is known about opioid mechanisms in invertebrates and protists.

1.1 Opioid Mechanisms in Invertebrate Metazoa

In contrast with data available for mammalian organisms, there is a paucity of information about opioid mechanisms in invertebrates. Opioid mechanisms and various opioid peptides have been reported in a mollusk, *Mytilus*, and an insect, *Leucophaea* (Stefano 1989). Granulocytes from these organisms respond to low opioid concentrations by adhering and clumping (Stefano 1989); these immunocytes have also been shown to produce endogenous opioids such as met-enkephalin (Stefano 1989). Interestingly enough, a synthetic analog of met-enkephalin has similar effects on both invertebrate and human immunocytes, in that it induces the synthesis of interleukin-1-like immunoreactivity (Stefano et al. 1991). Opioid mechanisms also seem to be present in worms: there is evidence for the presence of a pro-opiomelanocortin-like gene that codes for β-endorphin in the parasitic trematode *Schistosoma* (Duvaux-Miret et al. 1992).

[1]Biology Department, University of Puerto Rico, Río Piedras Campus, San Juan, PR 00931-3360, USA
[2]Department of Neuroscience, University of Pittsburgh, Pittsburgh, PA 15260, USA

It has been speculated that secretion of β-endorphin by this organism lowers the immune response of the host (Duvaux-Miret et al. 1992). Furthermore, opioid receptors similar in size to mammalian receptors have been reported in leech membranes by means of covalent binding of labeled β-endorphin (Zipser et al. 1988). All these data seem to suggest that opioid mechanisms in invertebrates may be similar to those present in more complex organisms.

1.2 Opioid Mechanisms in Unicellular Organisms

A provocative report has been published on the existence of an opioid mechanism in bacteria (Zagon and Mclaughlin 1992). Met-enkephalin inhibits the growth of *S. aureus* and other cells, whereas opioid antagonists accelerate replication. An opioid receptor has been detected in these microorganisms by means of specific binding experiments, and met-enkephalin has been detected by a radioimmunoassay in medium from cultures growing for 6 h. However, these findings need to be confirmed by rigorous pharmacological and biochemical data; for example, a number of opioid-like peptides have been reported that are not related to the opioid family in terms of amino acid sequence, such as deltorphins and casomorphin (Carr et al. 1990; Erspamer et al. 1989). Opioid effects on a protozoan cell were first reported in *Amoeba* (Jossefson and Johansson 1979), where morphine, met-enkephalin, and β-endorphin were shown to inhibit cation-induced pinocytosis. The effect was biphasic since only low concentrations were effective, giving rise to U-shaped dose-response curves. The inhibition is also naloxone-reversible, suggesting that it is receptor-mediated. Calcium ions may also be part of this opioid mechanism, since their addition abolished the inhibitory effect of opioids; this is similar to what has been reported in brain synaptosomes, where opioids inhibit calcium influx (Ueda et al. 1987). Unfortunately, no endogenous opioids have yet been reported in *Amoeba*, nor have signal transduction elements been characterized in this organism. An opioid mechanism has also been reported in the heterotrich ciliate *Stentor*: this cell responds to mechanical stimuli by contracting, and this response is inhibited by opioids, with β-endorphin being the most effective agonist (Marino and Wood 1993). The effect of β-endorphin is blocked by naloxone, which suggests that it is receptor-mediated. It is also blocked by prior treatment of the cells with pertussis toxin, suggesting the involvement of a G-protein. Furthermore, as in the case of *Tetrahymena* to be discussed below, a chronic exposure to the opioid results in an apparent tolerance. Voltage clamp analysis indicated that β-endorphin acts on *Stentor* by decreasing inward currents through the mechanoreceptor channel. Similar effects have been reported for the modulation by opioid agonists of calcium conductance in vertebrate systems (Seward et al. 1991), which suggests to the authors that mechanisms of signal transduction and neuromodulation originated early in evolution (Marino and Wood 1993). However, as in the case of *Amoeba* no endogenous opioids have been reported yet in *Stentor*, although, as will be seen in the next section, the *Tetrahymena* and the *Stentor* opiod mechanisms may be related.

2 The Opioid Mechanism in *Tetrahymena*

This is at the moment the best-characterized opioid mechanism in protists. A β-endorphin-like immunoreactivity has been reported in this organism (LeRoith et al. 1982), and we have recently confirmed it using affinity columns that exclude any possibility of artifacts due to sample contamination (Renaud et al. 1991). The original report included data that suggested that the β-endorphin-like material may be synthesized from a large precursor, proopiomelanocortin (POMC), that includes the sequence for adenocorticotropic hormone (ACTH). In addition, these investigators reported that this substance would bind to mammalian membranes in a manner similar to that of endogenous opioids (LeRoith et al. 1982). However, they offered no clue as to its function in *Tetrahymena*. Two possible functions have been postulated from recent literature: we have found that mammalian opioids and morphine inhibit phagocytosis in this ciliate, and since this effect is naloxone-reversible, it may be receptor-mediated (De Jesús and Renaud 1989). Other investigators have also reported that mammalian β-endorphin served as a chemoattractant for these cells and that this effect is also reversed by naloxone (O'Neill et al. 1988). Interestingly enough, the effective concentrations for chemotaxis were far lower than those that inhibit phagocytosis. It should also be noted that the effects on phagocytosis closely parallel those reported previously in *Amoeba*, since the effective concentrations are almost identical and the dose-response curves initially obtained were U-shaped because of the lack of response at the higher agonist concentrations. Therefore, the function of the endogenous opioid may be either to serve as chemoattractant or to modulate phagocytosis. Only experiments with the purified endogenous peptide will shed light on this question.

2.1 Pharmacological Characterization of the Endogenous Opioid

As stated previously, β-endorphin-like immunoreactivity has been detected in this ciliate (LeRoith et al. 1982), but it has yet to be characterized biochemically. The latter is a formidable task due to the low amounts of protein involved, which make purification and sequencing difficult; this question is probably more easily answered by recombinant DNA techniques. We are currently characterizing sequences from *Tetrahymena* genomic DNA that hybridize at low stringency with a mammalian POMC probe (Renaud et al. 1995). Nevertheless, even if the protein has yet to be purified and characterized, its biological activity may be studied in pharmacological experiments with partially purified preparations. *Tetrahymena* extracts were prepared from cell

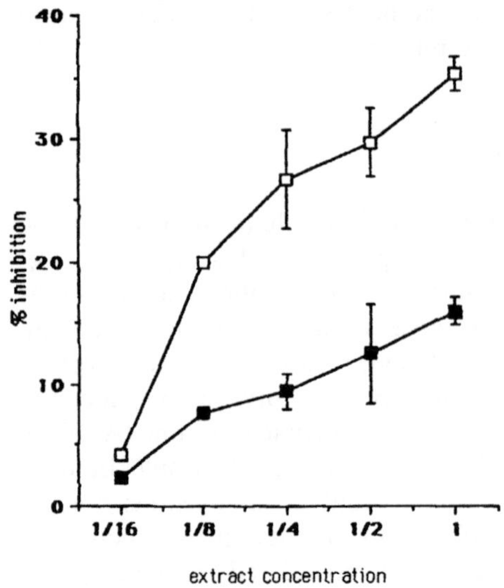

Fig. 1. Effect of partially purified β-endorphin-like immunoreactivity from *Tetrahymena* on phagocytosis. *Tetrahymena* cells grown for 48 h in defined medium (GIBCO) were homogenized with 1 N acetic acid, 0.1% 2-mercaptoethanol, supplemented with the following proteolytic inhibitors at 1 μg/ml: leupeptin, PMSF (phenylmethylsulfonyl fluoride), aprotinin, and pepstatin A. After centrifugation at 27 000 *g*, the supernatant was run through SepPack C_{18} columns, eluted first with 0.01 M trifluoroacetic acid (TFA), followed by 0.01 M TFA with 60% acetonitrile. This last eluate was dried, and the residue resuspended in phosphate buffer-saline (PBS) supplemented with proteolytic inhibitors, and the opioid purified by affinity chromatography using columns of anti-β-endorphin-Sepharose (Incstar, Clearwater, MN) according to the directions of the manufacturer. After elution of the opioid with 0.025 N HCl, the preparation was dried and stored at –70 °C until used in the phagocytosis assay. For this purpose, the extract was dissolved in 0.1 ml of the buffer used in the phagocytosis assay (De Jesús and Renaud 1988) and either added directly to a suspension of cells in buffer, or the extract solution was diluted prior to adding to the cells to one half, one third, one eighth, and one-sixteenth of the original concentration. At zero time the extract or its dilution was added together with the particles to be phagocytized (□); the volumes were adjusted so that the final volume of the cell suspension was 3.5 ml. The effect on phagocytosis was determined 6 min after the initiation of the process as the percent inhibition, calculated as described previously (Chiesa et al. 1993). These experiments were also carried out in the presence of both extract and 50 nM naloxone (■). In this and all the following figures the standard deviation of the mean is indicated by *vertical bars*

homogenates by hydrophobic and affinity chromatography, as described in the legend of Fig. 1. These partially purified preparations were then added to *Tetrahymena* cells, and its effect on phagocytosis scored, in both the presence and absence of naloxone. It can be seen in Fig. 1 that the extract will inhibit phagocytosis in *Tetrahymena* cells in a concentration-dependent manner. Furthermore, this effect is reversible by naloxone, suggesting that it is receptor-mediated. We do not know at this moment why the inhibition is not totally reversed with the high extract concentrations, but it may be surmised that a higher antagonist concentration would have been necessary to achieve total

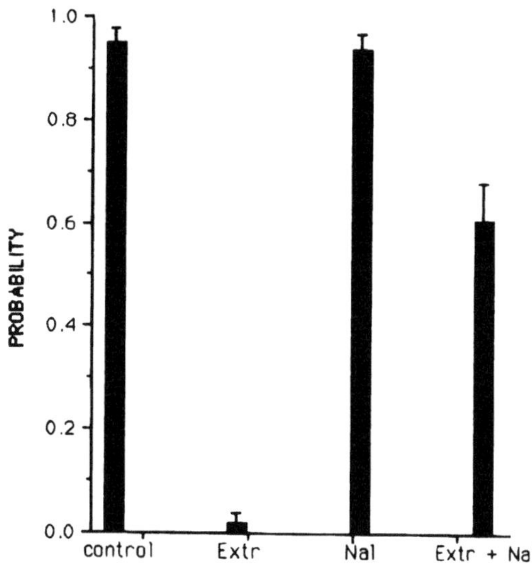

Fig. 2. Effect of partially purified β-endorphin-like immunoreactivity from *Tetrahymena* on the response to mechanical stimuli by *Stentor*. The probability of *Stentor* contracting when mechanically stimulated was tested in control cells and in cells in the presence of *Tetrahymena* extract *(Extr)*, or treated with 10 μM naloxone (*Nal*) and in cells treated with both *Tetrahymena* extract and naloxone (*Extr + Nal*). Naloxone-treated cells were exposed to this antagonist 45 min prior to the test. The extract was prepared as described in the legend to Fig. 1, and was dissolved in 1.0 ml of culture medium; 100 μl extract (10% total in 1.0 ml final volume) was added at time zero. In the case of controls and naloxone alone, 100 μl of culture medium was added at time zero; cells were tested in all cases after 5 min. A probability of 1.0 means that cells will contract 100% of the time, and a probability of 0.0 means that no cells will contract

reversion. It is of also of great interest to note that when this extract is tested on the *Stentor* system it has an effect similar to mammalian β-endorphin in that it inhibits mechanical stimulation-induced contraction, and this inhibition can be blocked by naloxone (Fig. 2). Thus, it seems that the endogenous opioid in *Tetrahymena* can be recognized as such by a different ciliate species; it remains to be established if it can also be recognized as an opioid by vertebrate cells although the work of LeRoith et al. suggests that it is so (LeRoith et al. 1982). This suggests that important elements of this opioid mechanism may have been conserved in evolution.

2.2 Pharmacological Characterization of the Opioid Receptor

We have used the modulation of phagocytosis by opioids as an assay system to study the pharmacological characteristics of the *Tetrahymena* opioid receptor, and have found several interesting properties. First of all, the receptor appears to desensitize rapidly in the presence of agonist, and this desensitization results in a biphasic dose-response curve (Fig. 3). However, a sustained, stronger in-

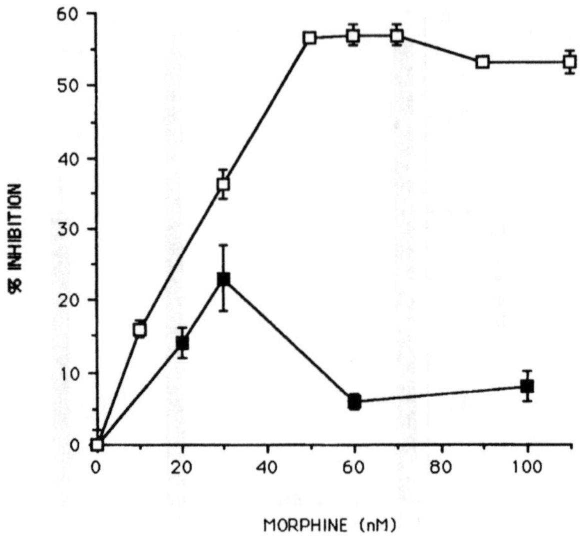

Fig. 3. Effect of morphine on phagocytosis in *Tetrahymena* when cells are preincubated for 3 min with the drug before the assay (■) and when the preincubation period is omitted (□); % inhibition was calculated as described previously (After Chiesa et al. 1993)

hibition is observed when the pre-incubation period with the agonist is abolished (Fig. 3). The resultant hyperbolic curves enable the establishment of a potency hierarchy (Table 1) that suggests that the receptor is mu-like due to its high affinity for morphine. Another mu-like characteristic is the affinity for naloxone which is comparable to that of mammalian mu-rich tissues (Chiesa et al. 1993).

Another important characteristic of the *Tetrahymena* opioid receptor is that in the presence of a chronic exposure (around 8 h) to morphine a state akin to tolerance is developed, since morphine no longer inhibits phagocytosis; some concentrations are actually stimulatory under these conditions (Fig. 4). Furthermore, if morphine is withdrawn form putatively tolerant cells by washing, an inhibition of phagocytosis is observed (Fig. 4). This suggests that a dependent state has been developed in which morphine is necessary for the cells

Table. 1. Potency (EC_{50}) and efficacy (E_{max}) of opioid agonists on phagocytosis in *Tetrahymena*. (After Chiesa et al. 1993)

Agonist	Receptor subtype	EC_{50}[a]	E_{max}(% inhibition)[b]
Morphine	μ	17.93 ± 1.58	54.7± 0.6
β-Endorphin	δ, ε	34.77 ± 2.6	51 ± 2.0
leu-Enkephalin	δ	21.73 ± 1.7	40 ± 6.2
met-Enkephalin	δ	28.07 ± 2.25	39.7± 2.1
Dynorphin$_{1\text{-}13}$	κ	18.53 ± 3.12	25.7± 0.6

[a]EC_{50} is defined as the agonist concentration that results in half the maximal inhibitory effect obtained with that particular agonist.
[b]E_{max} is defined as the maximal effect obtained with the agonist.

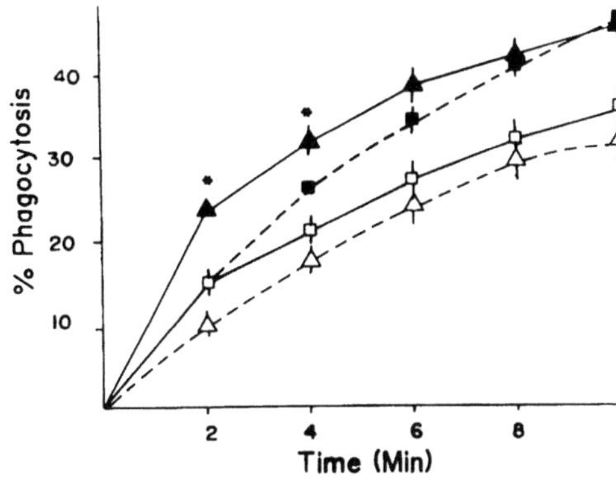

Fig. 4. Effect of a chronic exposure (16 h) to 30 nM morphine on phagocytosis in *Tetrahymena*. Morphine (30 nM) stimulates phagocytosis in chronically exposed, tolerant cells (▲); the *asterisks* indicate the time points in which the % phagocytosis is significantly higher ($p < 0.02$) than in control cells (■). When tolerant cells phagocytize in the absence of morphine, the phagocytic capacity is decreased (□) and is similar in magnitude to that of control cells in the presence of 30 nM morphine (△). (After Salamán et al. 1990)

to phagocytize at a basal level. This is further confirmed by the observation that when tolerant cells are preincubated with naloxone prior to the phagocytosis assay, the stimulatory effect of morphine on phagocytosis is abolished (Fig. 5). However, in the presence of both naloxone and a higher concentration of morphine (120 nM), an inhibition is observed that suggests that, as expected in the case of a true tolerant state, there has been a shift to the right of the dose-

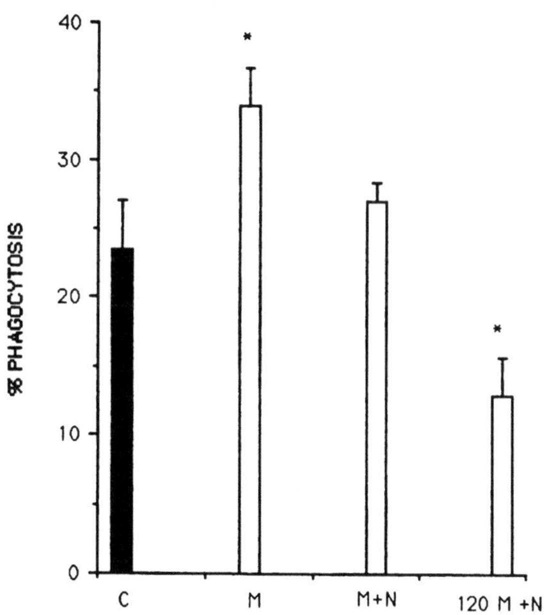

Fig. 5. Effect of naloxone on morphine modulation of phagocytosis in chronically exposed *Tetrahymena*. Cells cultured for 16 h in the presence of 30 nM morphine were incubated for 3 min in the presence of both 30 nM morphine and 30 nM naloxone (*M* + *N*); 120 nM morphine and 30 nM naloxone (*120 M* + *N*); and in the presence of 30 nM morphine only, with no antagonist present prior to the initiation of phagocytosis. We observe a stimulatory effect of morphine on phagocytosis ($M^* = P \leq 0.05$) when compared to control cells (*C*). However, 50 nM naloxone reverses this stimulation (M + N); and in the presence of 120 nM morphine and naloxone(120 M + N), a pronounced inhibitory effect is observed when compared to control cells ($P \leq 0.05$)

response curve (Fig. 5). Our observations on the effect of opioids on phagocytosis in *Tetrahymena* closely parallel the observed effect of opioids on murine peritoneal macrophages (Casellas et al. 1991). In the latter system, opioids inhibit phagocytosis by a dose-dependent, naloxone-reversible mechanism; and a chronic exposure will result in an apparent tolerance. Withdrawal of morphine from tolerant tissues has also been reported to inhibit secretion from tolerant enteric ganglia (Gintzler et al. 1987), which parallels the withdrawal-induced inhibition of phagocytosis which we have observed in tolerant *Tetrahymena* (Figs. 4, 5).

We have also tried to characterize the *Tetrahymena* opioid receptor by means of an irreversible ligand that binds specifically to the mu-subtype, *β*-funaltrexamine (Lee-Yuan et al. 1990). It can be observed that when *Tetrahymena* cells are incubated for 1 h at 25 °C in the presence of this compound, morphine can no longer inhibit phagocytosis in this organism (Fig. 6), which suggests that the receptor has been blocked by the antagonist. We have also attempted to label the opioid receptor in membrane fractions using ^{3}H-*β*-funaltrexamine; rat and mouse brain membrane fractions were included in these experiments as positive controls. In all cases, we observed two bands of

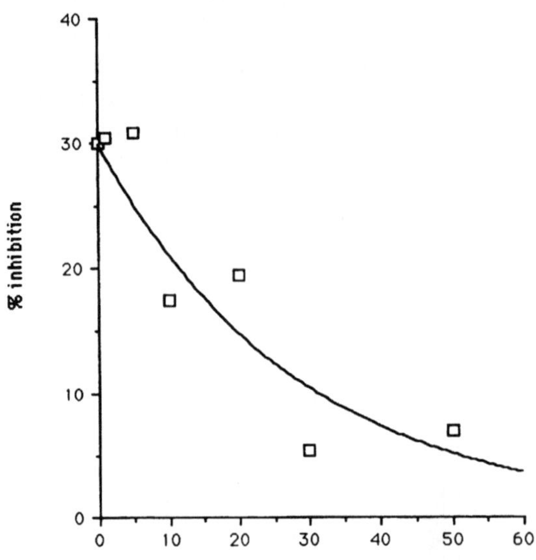

Fig. 6. Effect of pretreatment with *β*-funaltrexamine on morphine inhibition of phagocytosis in *Tetrahymena*. *Tetrahymena* cells were incubated in phagocytosis buffer in the presence of different concentrations of *β*-funaltrexamine with constant stirring at 25 °C; control cells were treated with an equivalent volume of methanol. After 30 min, cells were washed twice by centrifuging and resuspending in fresh buffer. The phagocytosis assay was performed as before (Chiesa et al. 1993) in the presence of 50 nM morphine. The results are illustrated in terms of % inhibition. It can be observed that concentrations of *β*-FNA between 1 and 5 nM had no effect on morphine inhibition of phagocytosis, whereas concentrations between 10 and 50 nM result in a dose-dependent abolition of this effect

approximately 53 and 56 kDa in the fluorography. However, the significance of this labeling is unknown because it was not abolished by preincubating with naloxone, as should have been the case for a genuine opioid receptor.

2.3 Biochemical Characterization of the Signal Transduction Pathway

We have previously presented evidence that the opioid mechanism in *Tetrahymena* involves G-proteins, since pertussis toxin treatment blocks the inhibitory effect of opioids (Fig. 7). We have also identified by means of Western blots the presence of Gα subunits of 51 and 59 kDa using an antibody specific for the conserved GTP-binding sequence (Renaud et al. 1995); these proteins await cloning and sequencing in order to establish their homology with other known G-proteins. However, there is so far no indication of the effector molecules that would be affected by the activated G-proteins, since we have

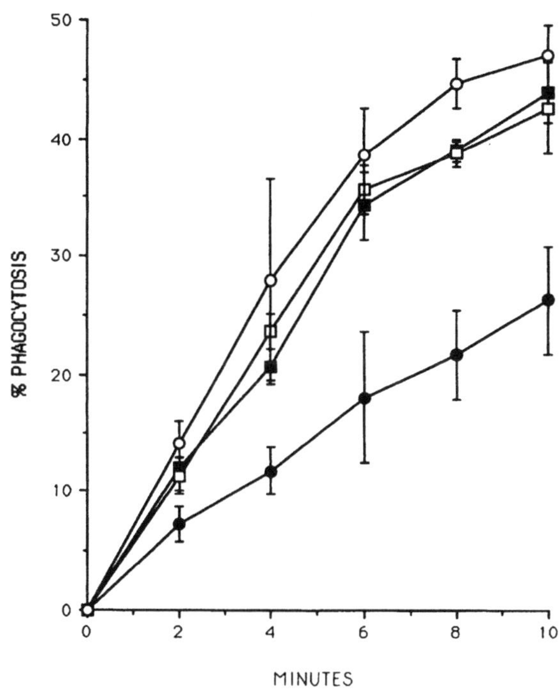

Fig. 7. Effect of pertussis toxin on met-enkephalin inhibition of phagocytosis in *Tetrahymena*. Cells were cultured in the presence and absence of 20 ng of pertussis toxin (PT) per ml for 2 h at room temperature, followed by washing by centrifugation and resuspension in phagocytosis buffer (De Jesús and Renaud 1989), and phagocytosis scored at different times in the presence and absence of 70 nM met-enkephalin. It may be seen that, although met-enkephalin inhibits phagocytosis in *Tetrahymena* by almost 50% (●) when compared to control cells (○), pretreatment of cells with PT abolishes this inhibitory effect (□); cells treated with PT but not with met-enkephalin (■). (After Renaud et al. 1991)

detected no changes in the levels of cyclic AMP or GMP in the presence of opiates or upon treatment with pertussis and cholera toxin or forskolin; similar results had been reported by other investigators (Kudo et al. 1985). The possibility remains that these proteins modulate ion channels, since we have found that cholera toxin stimulates phagocytosis in *Tetrahymena*, and calcium uptake during this process is also stimulated (Renaud et al. 1991).

3 Conclusions

A comparison of the data available for opioid mechanisms in *Tetrahymena* and various phyla strongly suggest the following important conclusions:

1. The effect of opioids in *Tetrahymena* closely parallels opioid effects on other cells such as *Amoeba, Stentor*, and murine macrophages.
2. The mechanism by which these effects take place on *Tetrahymena* probably involves elements similar to those involved in opioid signal transduction in metazoa, such as specific opioid receptors and G-proteins.
3. Opioids as signaling molecules appear to have originated early in evolution, prior to the development of metazoa.
4. Opioid mechanisms appear to have been conserved in evolution.

Acknowledgments. The author is grateful for the permission to use figures and tables from the Society of Protozoologists, and the journals Neuropeptides and Comparative Biochemistry and Physiology. The experiments done by the author and his coworkers have been supported by NSF grant MCB-9105024 and NIH grant S06 GM08102.

References

Carr RI, Webster D, Sadi D, Williams H (1990) Immunomodulation by opioids from dietary casein (exorphins). Ann NY Acad Sci 594: 374–376

Casellas AM, Guardiola H, Renaud FL (1991) Inhibition by opioids of phagocytosis in peritoneal macrophages. Neuropeptides 18: 35–40

Chiesa R, Silva WI, Renaud, FL (1993) Pharmacological characterization of an opioid receptor in the ciliate *Tetrahymena*. J Euk Microbiol 40: 800–804

Childers, SR (1991) Opioid receptor-coupled second messenger systems. Life Sci 48: 1991–2003

De Jesús S, Renaud FL (1989) Phagocytosis in *Tetrahymena thermophila*: naloxone-reversible inhibition by opiates. Comp Biochem Physiol 92C: 139–142

Duvaux-Miret O, Stefano GB, Smith EM, Dissous C, Capron A (1992) Immunosuppression in the definitive and intermediate hosts of the human parasite *Schistosoma mansoni* by release of immunoactive peptides. Proc Natl Acad Sci USA 89: 778–781

Erspamer V, Melchiorri P, Falconieri-Erspamer G, Negri L, Corsi R, Severini C, Barra D, Simmaco M, Kreil G (1989) Deltorphins: a family of naturally occuring peptides with high affinity and selectivity for δ-binding sites. Proc Natl Acad Sci USA 86: 5188–5192

Fischer EG (1988) Opioid peptides modulate immune functions. A review. Immunopharmacol Immunotoxicol 10: 265–326

Gintzler AR, Chan WC, Glass J (1987) Evoked release of methionine-enkephalin from tolerant/dependent enteric ganglia: paradoxical dependence on morphine. Proc Natl Acad Sci USA 88: 10178–10182

Jossefson JO, Johansson P (1979) Naloxone-reversible effect of opioids on pinocytosis in *Amoeba proteus*. Nature 282: 78–80

Kudo S, Muto Y, Nozawa Y (1985) Regulation by calcium of hormone-insensitive adenylate cyclase and calmodulin-dependent guanylate cyclase in *Tetrahymena* plasma membrane. Comp Biochem Physiol 80B: 813–816

Laugwitz K-L, Offermanns S, Spicher K, Schultz G (1993) μ and δ opioid receptors differentially couple to G protein subtypes in membranes of human neuroblastoma SH-SY5Y cells. Neuron 10: 233–242

Lee-Yuan L-C, Shuixing L, Tallarida RJ (1990) Studies on kinetics of [^{3}H]β-funaltrexamine binding to μ opioid receptor. Mol Pharmacol 37: 243–250

LeRoith D, Liotta S, Roth J, Shiloach J, Lewis ME, Pert CB, Krieger DT (1982) Corticotropin and β-endorphin-like materials are native to unicellular organisms. Proc Natl Acad Sci USA 79: 2086–2090

Marino MJ, Wood DC (1993) β-endorphin modulates a mechanoreceptor channel in the protozoan *Stentor*. Comp Physiol A 173: 233–240

North RA, Williams JT, Surprenant A, Christie MJ (1987) μ and δ receptors belong to a family of receptors that are coupled to potassium channels. Proc Natl Acad Sci USA 84: 5487–5491

O'Neill B, Pert CB, Ruff RM, Smith CC, Higgins WJ, Zipser B (1988) Identification and initial characterization of the opiate receptor in the ciliated protozan *Tetrahymena*. Brain Res 450: 303–315

Renaud FL, Chiesa R, De Jesús JM, López A, Miranda J, Tomassini N (1991) Hormones and signal transduction in protozoa. Comp Biochem Physiol 100A: 41–45

Renaud FL, Colón I, Lebrón J, Ortiz N, Rodríguez F, Cadilla C (1995) A novel opiod mechanism seems to modulate phagocytosis in *Tetrahymena*. J Euk Microbiol 42: 205–207

Salamán A, Román M, Renaud FL, Silva WI (1990) Effect of chronic opioid treatment on phagocytosis in *Tetrahymena*. Neuropeptides 16: 115–120

Seward E, Hammond C, Henderson G (1991) μ-opioid receptor-mediated inhibition of the N-type calcium channel current. Proc R Soc Lond B 244: 129–135

Simon EJ, Hiller JM (1994) Opioid peptides and opiod receptors. In: Siegel GJ, Agranoff BW, Albers RW, Molinoff PB (eds) Basic neurochemistry: molecular, cellular and medical aspects, Chapt 15, 5th Ed. Raven Press, New York, pp 321–339

Stefano GB (1989) Role of opioid neuropeptides in immunoregulation. Prog Neurobiol 33: 149–159

Stefano GB, Smith EM, Hughes TK (1991) Opioid induction of immunoreactive interleukin-1 in *Mytilus edulis* and human immunocytes: an interleukin-1-like substance in invertebrate neural tissue. J Neuroimmunol 32: 29–34

Ueda H, Tamura S, Satoh M, Takagi H (1987) Excess release of substance P from the spinal cord of mice during morphine withdrawal and involvement of the enhancement of presynaptic Ca^{2+} entry. Brain Res 425: 101–105

Zagon IS, McLaughlin PJ (1992) An opioid growth factor regulates replication of microorganisms. Life Sci 50: 1179–1187

Zipser B, O'Neill JB, Ruff MR, Smith CC, Higgins WJ, Pert CB (1988) The opiate receptor: a single 110 kDa recognition molecule appears to be conserved in *Tetrahymena*, leech and rat. Brain Res 463: 296–304

Adenylate and Guanylate Cyclases in *Tetrahymena*

S. UMEKI[1] and Y. NOZAWA[2]

1 Introduction

Cyclic nucleotides, such as cyclic AMP (cAMP) and cyclic GMP (cGMP), are of fundamental importance in regulating the many physiologic processes in a wide variety of living cells, acting as messengers interacting between receptor sites on the cell membrane and intracellular activity. Both adenylate cyclase (A-cyclase) and guanylate cyclase (G-cyclase), which catalyze the formation of cyclic nucleotides, are known to exist in most organisms. A free-living unicellular eukaryote, *Tetrahymena*, represents a suitable cell system for studies of the regulation of intracellular events by cyclic nucleotides (Nozawa and Thompson 1971), since it contains high activities of cyclic nucleotide phosphodiesterase (PDE; Dickinson et al. 1977; Kudo et al. 1980), A-cyclase (Rozensweig and Kindler 1972; Voichick et al. 1973), G-cyclase Nagao et al. 1979; Nakazawa et al. 1979), cyclic nucleotide-dependent protein kinases (Murofushi 1974), and calmodulin (CaM; Kakiuchi et al. 1981; Yazawa et al. 1981), which are involved in the metabolism of cyclic nucleotides. This chapter will dwell on the regulatory mechanisms in the metabolism of *Tetrahymena* cyclic nucleotides involved in the two cyclases, particularly as they relate to the molecular mechanisms of regulation of those enzymes and CaM, and will discuss the structural features of *Tetrahymena* CaM which endow it with its special functional capabilities, and its intracellular distribution.

2 Cyclic Nucleotide Metabolism in *Tetrahymena*

2.1 Cell Growth- and Cycle-Associated Changes

Modulation of cAMP and cGMP is associated with several physiologic processes in *Tetrahymena*, including cell growth and cycling. Changes in cyclic nucleotide concentrations after exposure to various agonists have been char-

[1]Department of Medicine, Toshida-kai Kumeda Hospital, 2944 Obu-Cho, Kishiwada, Osaka 596, Japan
[2]Department of Biochemistry, Gifu University School of Medicine, 40 Tsukasamachi, Gifu 500, Japan

acterized with radioimmunoassays of cyclic nucleotides extracted from whole cells. Although cAMP has been shown to be present in the whole cell of *Tetrahymena* (Kuno et al. 1981), high levels of cAMP have been observed in an early exponential phase of the growth cycle (Voichick et al. 1973; Kariya et al. 1974), being associated with elevated A-cyclase and decreased cAMP PDE activities (Voichick et al. 1973; Shimonaka and Nozawa 1977). Kudo et al. (1981a) studied growth-related changes in the activities of cyclic nucleotide-metabolizing enzymes in the *Tetrahymena* whole cell (Fig. 1). Although the variations in the activities of A-cyclase and cAMP PDE were not significant during the growth of cultures, G-cyclase activity and CaM content showed a maximum at the middle exponential phase followed by a minimum at the early stationary phase and a subsequent rise (see the Sect. 5 for CaM). The activity of cGMP PDE temporarily increased only in the middle exponential phase. These results suggest a parallel change between G-cyclase activity and cGMP content and a close relationship between G-cyclase and CaM in the cell growth cycle of *Tetrahymena*.

Dickinson et al. (1976) investigated cAMP, A-cyclase, and cAMP PDE activities in the natural cell cycle of selection-synchronized *Tetrahymena* (Fig. 2). A large increase in intracellular cAMP level concomitant with cyto-

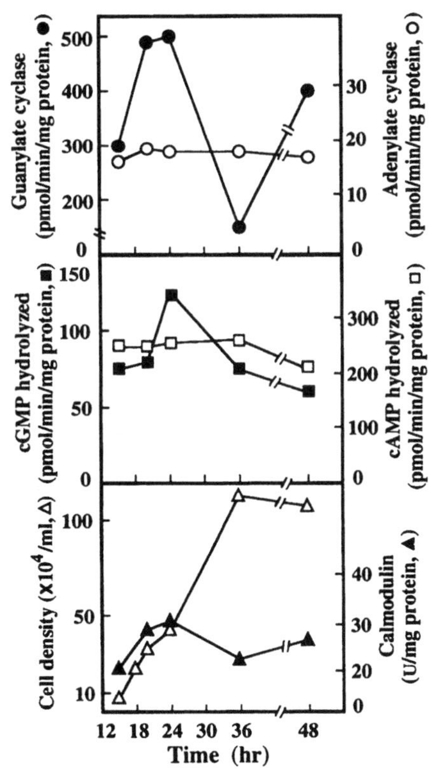

Fig. 1. Growth-related changes in cyclasses and cyclic nucleotide PDE activities and calmodulin content in *Tetrahymena* whole cell. *Upper panel* A-cyclase (○) and G-cyclase (●) activities; *middle panel* cAMP PDE (□) and cGMP PDE (■) activities; lower panel cell density (△) and calmodulin content (▲). (Kudo et al. 1981a)

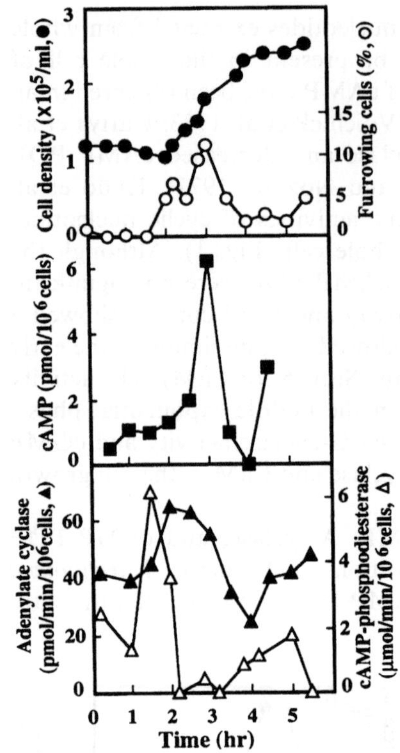

Fig. 2. Cell cycle-related changes in cAMP content and A-cyclase and cAMP PDE activities in *Tetrahymena* whole cell. *Upper panel* Cell density (●) and percentage dividing cells (○); *middle panel* cAMP content (■); *lower panel* A-cyclase (▲) and cAMP PDE (△) activities. (Dickinson et al. 1976)

kinesis was observed. Shortly after division, the cAMP content quickly diminished. High cAMP levels may be produced by an increased A-cyclase activity coincident with decreased cAMP PDE activity during cell division. Kariya et al. (1974) demonstrated, in synchronized cultures of *Tetrahymena*, that dibutyryl cAMP inhibited protein and RNA syntheses in the G_1 phase and the subsequent DNA synthesis in the S phase, suggesting that cAMP plays an important role in regulating initiation from the G_1 phase to the S phase. However, Kassis and Zeuthen (1979) reported maximum cAMP levels at the end of the S phase and minimum cAMP levels at division in *Tetrahymena*, synchronized with the one thermal shock per generation method. The exact reasons for the discrepancies are unknown, but they may be a reflection of the different synchrony methods used in the two studies. During the free-running cell cycle in the interval of time between the first and second synchronous division, Zimmerman et al. (1981) demonstrated that cAMP content reached a high level in the G_2 phase and peaked at division.

Cell cycle-associated changes in the intracellular content of cGMP, which exists along the entire plasma membrane and within the cilia of *Tetrahymena* (Kohidai et al. 1992), have been investigated. Using a single hypoxia-synchronized *Tetrahymena*, Gray et al. (1977) showed that cGMP fell to a minimum before the first synchronous division and then reached a peak in relation

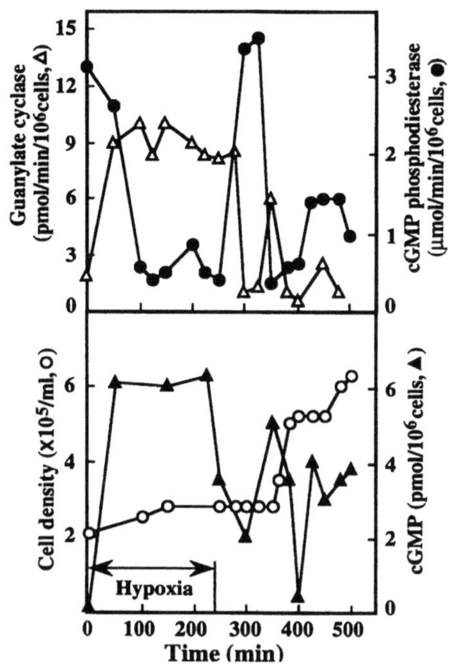

Fig. 3. Cell cycle-related changes in cGMP content and G-cyclase, and cGMP PDE activities in *Tetrahymena* whole cell. *Upper panel* G-cyclase (△) and cGMP PDE (●) activities; *lower panel* Cell density (○) and cGMP content (▲). (Gray et al. 1977)

to the division (Fig. 3). The increase in the cGMP level was associated with an increase in G-cyclase activity coincident with decreased cGMP PDE activity. On the other hand, Kudo et al. (1981b) found, in the heat-shock-synchronized *Tetrahymena*, that G-cyclase activity fell to a minimum at the first synchronous division and then quickly rose toward the S phase, being parallel to changes in CaM content during the cell cycle (Fig. 4). These results suggest that fluctuations in G-cyclase activity during the cell cycle are dependent on the con-

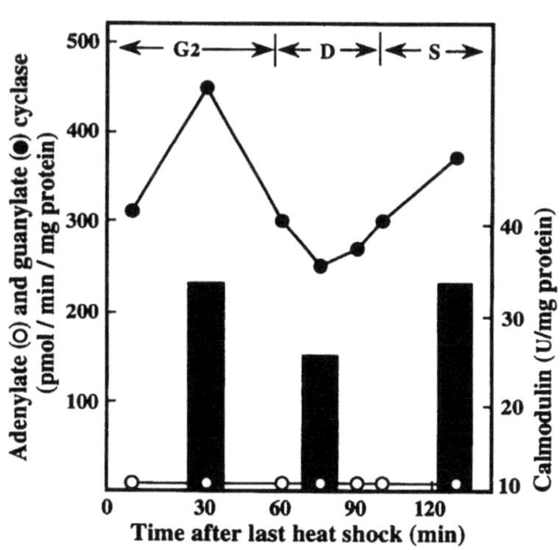

Fig. 4. Cell cycle-related changes in A- and G-cyclase activities and calmodulin content in the *Tetrahymena* whole cell. ○ A-cyclase activity; ● G-cyclase activity. *Column* indicates calmodulin content. (Kudo et al. 1981b)

centration of CaM. It has been considered that cGMP acts antagonistically to cAMP in a variety of cell activities, and that an increase in cGMP content promotes cell growth (Goldberg et al. 1973; Seifert and Rudland 1974). In *Tetrahymena*, however, the above-mentioned cAMP and cGMP findings do not seem to be in accordance with these ideas.

2.2 Involvement in Biological Regulation

The nature of the stimulus that affects cyclic nucleotide formation has been gradually disclosed. Cyclic nucleotides are known to be intracellular signaling molecules associated with ion channels. Nandini-Kishore and Thompson (1979) observed that glucose stimulated cAMP formation via a depolarization-gated Ca^{2+} influx followed by an apparent release of cellular epinephrine. This finding is supported by circumstantial evidence indicating that *Tetrahymena* is able to synthesize epinephrine (Janakidevi et al. 1966; Gunderson and Thompson 1983) and that it responds to it by increasing cAMP levels through a β-type adrenergic receptor coupled to A-cyclase (Rozensweig and Kindler 1972; Umeki and Nozawa 1984) and through the reduction of cAMP PDE (Umeki and Nozawa 1985). In this regard, an increase in the Na^+ entry produced by exogenous glucose in *Tetrahymena* (Aomine 1976) may well be sufficient to depolarize the membrane, thereby opening the Ca^{2+} gates to an even greater influx of this critical divalent cation (Eckert 1972).

The effects of hyperpolarizing changes in the concentration of additional K^+ ion concomitant with Ca^{2+} ion on the regulation of cAMP formation in *Tetrahymena* have also been investigated. Schultz and Schönborn (1994) reported that hyperpolarization of *Tetrahymena* by dilution of extracellular K^+ concentration stimulated intracellular cAMP generation in a dose-response-like manner correlated with changes in external K^+ concentrations. In addition, they found that three K^+ channel blockers, tetraethylammonium, Cs^+, and quinine, dose-dependently inhibited cAMP biosynthesis, and that an elevation of the Ca^{2+} ion, which causes a hyperpolarization-induced fast-swimming response, also elicited cAMP generation. These results indicate that the K^+ resting conductance and the capacity for cAMP generation may be coregulated. During various environmental changes, *Tetrahymena* appears to actively regulate its resting conductances, mainly carried by K^+, probably through controlling intracellular cAMP formation, to maintain its membrane potential and intracellular ion composition within narrow limits.

On the other hand, profound changes in the microsomal fatty acyl-CoA desaturase activities and cAMP generation of *Tetrahymena* originally grown in the glucose-deficient medium, were observed following the administration of glucose or β-adrenergic agonists (Fig. 5; Umeki and Nozawa 1985). There was a great increase in stearoyl-CoA desaturase activity coincident with a two-fold decrease in oleoyl-CoA desaturase activity over the first 2 h after the administration of these compounds. These results indicate that the increased cAMP

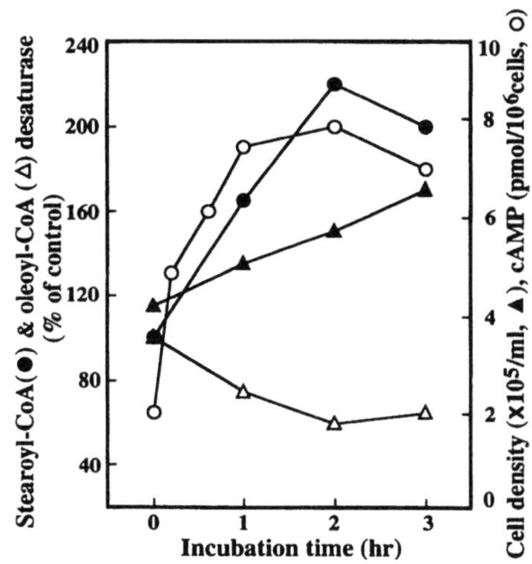

Fig. 5. Relationships between the cAMP accumulation and modification of stearoyl-CoA and oleoyl-CoA desaturase activities in epinephrine-treated *Tetrahymena*. This figure shows the time course of each factor after the treatment with theophylline and epinephrine. ▲ Cell density; ○ cAMP content; ● stearoyl-CoA desaturase acitivity; △ oleoyl-CoA desaturase activity

due to glucose and β-adrenergic agonists may modulate the microsomal fatty acyl-CoA desaturation system, which leads to the adjustment of the membrane lipid composition to maintain an adequate membrane fluidity required for eliciting proper membrane functions (Nozawa and Umeki 1988).

3 Cyclases Involved in Cell Metabolism and Functions

The regulation of A- and G-cyclases is of great interest, since these enzymes may constitute the initial intracellular target molecules for excitatory signal transduction, which finally are connected to such physiological processes as alteration in swimming speed and direction, regulation of ion fluxes, desensitization, and adaptation in protozoa. In considering a physiological function for the cyclic nucleotides in *Tetrahymena*, a precise subcellular localization of cyclases is of substantial help. Subcellular localization of both cyclases has been investigated for *Tetrahymena* (Table 1; Watanabe and Nozawa 1982). A- and G-cyclases of high specific activities were localized in the surface membrane pellicles from *Tetrahymena*. These results suggest that the two cyclases are substantially associated with particulate fractions, unlike the subcellular distribution in sea urchin sperm (Garbers and Gray 1974) and *Physarum polycephalum* (Lovely and Threlfall 1979), where the activity is mainly in the cytosol.

Shimonaka and Nozawa (1977) studied the effects of a temperature-induced phase transition of *Tetrahymena* membrane lipids on the activity of plasma membrane-bound A-cyclase, and found that discontinuities in the Arrhenius plots of A-cyclase activity, primarily associated with the surface membrane, occurred at temperatures at which membrane lipids undergo a phase transition producing changes in the physical state. In general, it is well known that the dynamism of the biological membrane is closely involved in both cyclases

Table 1. Intracellular distribution of A- and G-cyclases in *Tetrahymena*. (Watanabe and Nozawa 1982)

Subfractions	Protein[a]	A-cyclase		G-cyclase	
		Total activity[b]	Specific activity[c]	Total activity[b]	Specific actvity[c]
Cilia	5.6	0	0	176 (1)[d]	31
Pellicles	20.3	333 (89)	163	13688 (81)	673
Mitochondria	28.4	508 (12)	18	2462 (15)	87
Microsomes	24.6	352 (8)	14	683 (4)	28
Postmicrosomal supernatant	70.2	0	0	0	0

[a]mg/10^8 cells.
[b]pmol/min.
[c]pmol/min/mg protein.

biosynthesizing cAMP and cGMP. These results indicate a close interaction between the enzyme activity and biological membrane fluidity, which is closely associated with maintaining the optimal level for various metabolic functions of the cell.

Precise recognition of environmental influences and a quick, appropriate response to them are vitally pivotal for free-living unicellular organisms. In *Tetrahymena*, the A-cyclase/cAMP system plays a role in sexual recognition through chemoreceptorial detection of the pheromones released by different mating types (O'Day 1981). A-cyclase purified from *Paramecium*, on the other hand, when incorporated into an artificial lipid bilayer membrane, revealed properties of a voltage-independent K^+ channel (Schultz et al. 1992). This indicates that, in *Paramecium*, the transmembrane enzyme A-cyclase itself may secondarily operate as a carrier of the K^+ resting conductance.

Schultz et al. (1983) noted that the ciliary membrane from *Tetrahymena* contained a high specific activity of G-cyclase and a discrete Ca^{2+}-permeability, as demonstrated by Ca^{2+} flux measurements using the Ca^{2+} indicator dye arsenazo III. In analogy with the excitable ciliary membrane of the larger relative *Paramecium* (Schultz et al. 1986), *Tetrahymena* G-cyclase is thus involved in ciliary beating of the cell and appears to be regulated by a voltage-gated inward Ca^{2+} current.

4 Regulatory Mechanisms of Cyclases

4.1 Adenylate Cyclase

Dozens of hormones and neurotransmitters have been found to either stimulate or inhibit A-cyclase (catalytic subunit) by pathways that consist of two other distinct types of plasma membrane-associated proteins; i.e., hormone receptors and guanine nucleotide-binding regulatory proteins (G proteins), which mediate hormonal activation. The binding of hormone to a stimulatory receptor at the plasma membrane triggers signal-transducing G_s activation followed by A-cyclase activation. This is supported by hormone-stimulated cAMP generation observed in a system reconstituted with homogeneous β-adrenergic receptors and G_s in phospholipid vesicles (May et al. 1995).

In *Tetrahymena*, unlike a mammalian and other eukaryotic cells, the activation mechanisms of A-cyclase with regard to hormone receptors on the plasma membrane and receptor-A-cyclase coupling have not been characterized in detail. Rozensweig and Kindler (1972) demonstrated that *Tetrahymena* contained A-cyclase responsive to mammalian A-cyclase activators, such as epinephrine, serotonin, and fluoride. Later, they found that dispersion of the membranes with Triton X-100 led to a purified preparation of an epinephrine-insensitive enzyme (Kassis and Kindler 1975). However, Kudo et al. (1985) reported that *Tetrahymena* A-cyclase was not responsive to catecholamines, fluoride, guanine nucleotie, and forskolin (a lipid-soluble diterpene), which activates eukaryotic A-cyclases via interaction with certain hydrophobic domains of the A-cyclase system other than the catalytic subunit. Although this discrepancy is unclear, Kudo et al. (1985) suggest that *Tetrahymena* A-cyclase has a catalytic subunit functionally lacking the G protein and hormone receptors. There is no evidence that the A-cyclase is coupled to G proteins in *Paramecium* either (Schultz et al. 1987).

In analogy to some A-cyclase preparations devoid of a functional G protein (Ross 1982), *Tetrahymena* enzyme is also activated much more by Mn^{2+} than Mg^{2+} (Nakazawa et al. 1979). In addition, *Tetrahymena* A-cyclase activity has been reported to increase at a low free Ca^{2+} level, decrease at high levels, and to not respond to CaM (Table 2; Nagao et al. 1979; Watanabe and Nozawa 1982), indicating that low intracellular Ca^{2+} concentrations are required for the maximal activity of the enzyme. Similar changes have also been observed in *Paramecium* A-cyclase (Schultz et al. 1987). In this respect, Ca^{2+}/Mg^{2+}-ATPase, which exists in the microsomes as a Ca^{2+}-pump, appears to play a crucial

Table 2. Effects of Ca^{2+} and *Tetrahymena* calmodulin on *Tetrahymena* A- and G-cyclase activities. (Nagao et al. 1979; Watanabe and Nozawa 1982)

Additions	A-cyclase (pmol/min/mg protein)	G-cyclase (pmol/min/mg protein)
EGTA (0.1 mM)	62.5	2.3
Ca^{2+} (0.2 mM)	23.4	31.0
EGTA (0.1 mM) + CaM	53.2	2.7
Ca^{2+} (0.2 mM) + CaM	20.6	551.6
Ca^{2+} (0.2 mM) + CaM + TFP[a] (25 μM)	20.4	363.0
Ca^{2+} (0.2 mM) + CaM + TFP (50 μM)	19.5	66.9

[a]Trifluoperazine.

role in lowering the free intracelluler Ca^{2+} level, being accompanied by microsomal ATP-dependent Ca^{2+} uptake (Muto and Nozawa 1984). In addition, in *Tetrahymena* and *Paramecium*, A-cyclase is characterized by not being controlled through the Ca^{2+}/CaM system which produces G-cyclase activation (Nagao et al. 1979; Schultz and Klumpp 1983). In the excitable ciliary membrane from *Paramecium*, K^+ was found to activate A-cyclase at concentrations around 20 mM, at which its stimulatory potency was dependent on the free Mg^{2+} concentration (Schultz et al. 1987).

It has been hypothesized that A-cyclase has a dual life, serving as an ion channel as well as a transporter (Krupinski et al. 1989), on the basis of the unique topographical homology between A-cyclase and various ion channels and transporters. In *Paramecium*, of interest, A-cyclase has been found to exhibit pore-forming activity with different selectivities for cations, K^+, Na^+, Ca^{2+}, Mg^{2+}, Cs^+, and Li^+ (Schultz et al. 1992). Kameyama et al. (1982), in investigating phospholipase A_2 action on *Tetrahymena* A-cyclase, found that in the plasma membranes in which 45% of the total phosphatidylethanolamine was converted to its lysolipid by phospholipase A_2, the reduction in A-cyclase activity was small. These results tempt us to consider that the plasma membrane A-cyclase may not appreciably respond to its lipid environment associated with membrane fluidity. The properties of both the A- and G-cyclases of *Tetrahymena* are summarized in Table 3.

4.2 Guanylate Cyclase

Different species of G-cyclase, which appear to exist in all organisms, are often distinguished by their localization in various cellular compartments containing the plasma membrane, cytoplasma, and a detergent-insoluble cytoskeletal fraction. It is possible to classify G-cyclase into two groups based on their general structure and cellular distribution. The first group is the particulate enzyme, which contains a second subgroup of G-cyclases that are activated by the Ca^{2+}/CaM system, as observed in the ciliary membranes of the protozoans *Tetrahymena* and *Paramecium* (Kakiuchi et al. 1981; Schultz et al. 1986). The second group is considered to exist as heterodimeric forms containing heme as a prosthetic group, and is activated by the vasodilatory agents nitroprusside and nitroglycerin (Wong and Garbers 1992).

The subcellular distribution of *Tetrahymena* G-cyclase after cell disruption is almost totally particulate principally occupied by the surface membrane pellicles (Table 1). More than 80% of the total or specific activity of G-cyclase was found in the pellicular membrane, resembling results obtained with the plasma membrane enzymes of other cells. In *Tetrahymena* and *Paramecium*, significant G-cyclase activity in soluble fractions has still not been determined using ordinary methods.

The optimal concentrations of Tetrahymena G-cyclase for cations Mg^{2+} and Mn^{2+} were 3 and 1 mM, respectively (Table 3; Nakazawa et al. 1979); the

Table 3. Characterization of *Tetrahymena* A- and G-cyclases. (Nagao et al. 1979; Nakazawa et al. 1979; Kudo et al. 1985)

Factors	A-cyclase	G-cyclase
K_m for substrate (mM)	0.25	0.66
pH optimum	11.5	11.5
Optimal conc for cation (mM)		
Mg^{2+}	10	3
Mn^{2+}	1	1
Effectors		
Mg^{2+} (3 mM)	Activated	Activated
Mn^{2+} (1 mM)	Activated	Activated
Ca^{2+} (1 mM)	Inhibited	Activated
Co^{2+} (1 mM)	Activated	Activated
Epinephrine (0.1 mM)	Unresponsive	–[b]
Serotonin (0.1 mM)	Unresponsive	–
NaF (10 mM)	Inhibited	–
GTP (1 mM)	Unresponsive	Unresponsive
Forskolin (0.05 mM)	Unresponsive	–
EGTA (0.125 mM)	Activated	Unresponsive
NaN_3 (1 mM)	–	Unresponsive
Sodium nitroprusside (1 mM)	–	Inhibited
MNNG[a] (1 mM)	–	Unresponsive
$NaNo_2$ (20 mM)	–	Inhibited
Cytochrome (5 μg)	–	Unresponsive
Chymotrypsinogen (5 μg)	–	Unresponsive
Ovalbumin (5 μg)	–	Unresponsive
Bovine serum albumin (5 μg)	–	Unresponsive
Troponin-C (5 μg)	–	Unresponsive
Triton X-100 (0.1%)	–	Inhibited
Lubrol PX (0.1%)	–	Inhibited
Ca^{2+}/CaM	Unresponsive	Activated

[a]N-methyl-N′-nitro-N-nitrosoguanidine.
[b]Not determined.

former produced twofold higher enzyme activity in the absence of Ca^{2+}/CaM than did the latter. Similar changes were observed in *Paramecium* G-cyclase (Klumpp and Schultz 1982). In general, the G-cyclases in microorganisms and mammalian cells require Mg^{2+} or the more preferred cation Mn^{2+} as a cofactor for full activity. However, the cyclase activity in human peripheral blood lymphocytes is as high in the presence of Mg^{2+} as in that of Mn^{2+} (Coffey et al. 1978). These results indicate that Mg^{2+} probably is a more physiological cofactor for G-cyclase than Mn^{2+}, since the former is generally present in cells in a 100 times greater amount than the latter. The effects of other divalent cations on the enzyme activity in the absence of CaM have varied from being only partially effective (Ca^{2+} and Co^{2+}) to being noneffective (Fe^{2+}, Cu^{2+}, and Zn^{2+}). Metals such as Cu^{2+} and Zn^{2+} are, as a rule, considered to be inhibitory (Böhme et al. 1974).

To clarify the interactions between the Ca^{2+}/CaM system and Mg^{2+} or Mn^{2+}, Kudo et al. (1982a) determined the G-cyclase activity in *Tetrahymena* pellicular membranes as a function of the Mg^{2+} or Mn^{2+} concentration in the absence or presence of CaM (Fig. 6). Both basal and CaM-stimulated G-cyclase required the divalent cation for activation. In the presence of Ca^{2+} minus CaM, the basal enzyme activity was two times higher with various concentrations of Mn^{2+} than it was with ones of Mg^{2+}, contrasting with the results in the absence of Ca^{2+}/CaM (Nakazawa et al. 1979). The addition of CaM produced a considerable amount of Mg^{2+}-or Mn^{2+}-dependent enzyme activity. The Mn^{2+}-dependent activity, even when EGTA was substituted for Ca^{2+}, was as great as that stimulated by CaM in the presence of Ca^{2+}. Under the same conditions, however, the Mg^{2+}-dependent CaM-stimulated activity essentially required Ca^{2+}. In this regard, it is suggested that the affinity of G-cyclase for either Mg^{2+} or Mn^{2+} is decreased by adding CaM, due to the double reciprocal and Hill plots of the enzyme activity versus the factor of Mg^{2+} or Mn^{2+} (Kudo et al. 1982a). Additionally, it has been found that enzyme activity with an optimal concentration of Mg^{2+} is lower than that with the same concentration of Mn^{2+} at lower CaM concentrations. However, at higher CaM concentrations, the former exceeds the latter. On the other hand, even in the presence of Mg^{2+} and CaM, Ca^{2+} concentrations of more than 100 μM have been observed to reduce enzyme activity (Nagao et al. 1981a), possibly because of competition for the sites at which Mg^{2+} enhances the reactivity of the catalytic site or because of Ca^{2+} binding to distinct, allosteric inhibitory sites. Only the G-cyclase from *Paramecium*, a genus related to *Tetrahymena*, has been demonstrated to be sensitive to similar concentrations of Ca^{2+} (Klumpp and Schultz 1982).

In the protozoa *Tetrahymena* and *Paramecium*, CaM mediates the stimulation of particulate G-cyclase by Ca^{2+} (Kakiuchi et al. 1981; Schultz et al. 1986). The effect of the Ca^{2+}/CaM system on *Tetrahymena* G-cyclase activity

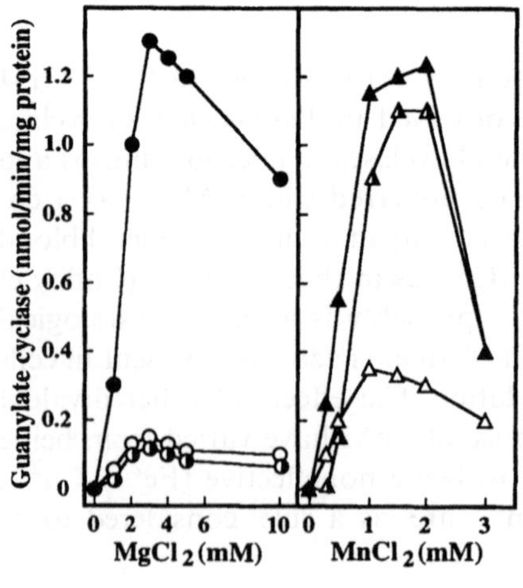

Fig. 6. Effects of Mg^{2+} and Mn^{2+} on *Tetrahymena* G-cyclase activity. *Left panel* Mg^{2+}; *right panel* Mn^{2+}. The assays were done in the presence of 0.1 mM Ca^{2+} (○, △), 0.1 mM Ca^{2+} plus 1 μg of calmodulin (●, ▲), or 0.2 mM EGTA plus 1 μg of calmodulin (◑, ◭). (Kudo et al. 1982a)

is described in Table 2 (Nagao et al. 1979). *Tetrahymena* CaM had no stimulatory effect on enzyme activity when EGTA was used in place of Ca^{2+}. Once Ca^{2+} was added, CaM brought about a striking increase in the cyclase activity. In analogy to *Tetrahymena* CaM, isolated *Paramecium* protein also activated *Tetrahymena* G-cyclase in a Ca^{2+}-dependent fashion in the presence of Mg^{2+} and Ca^{2+} (Kudo et al. 1981c), prompting us to suggest they have a close evolutionary relationship. Of some proteins, including a Ca^{2+}-binding protein troponin C from rabbit skeletal muscle and CaMs from bovine brain, *Renilla*, *Chlamydomonas*, *Dictyostelium*, spinach and English cucumber, however, none mimicked the stimulation of the enzyme mediated by *Tetrahymena* CaM (Table 3; Nagao et al. 1979; Kudo et al. 1982b). A particularly specific feature of *Tetrahymena* CaM is its ability to activate G-cyclase. Postulating that mammalian CaMs have lost activating G-cyclase during the course of evolution, it can be envisioned that *Tetrahymena* CaM is an ancestral form of mammalian CaMs.

Trifluoperazine (TFP) and chlorpromazine, phenothiazine psychotropic agents which have been known to prevent the stimulation of many cyclic nucleotide-metabolizing enzymes by binding to CaM (Weiss and Levin 1978) in the presence of Ca^{2+}, dose-dependently inhibited the Ca^{2+}/CaM-stimulated activity of *Tetrahymena* G-cyclase (Table 2 and Nagao et al. 1981a). The former inhibited the Ca^{2+}/CaM-stimulated activity of G-cyclase by approximately 50% over increasing ranges of Ca^{2+}. This implies that even greater concentrations of Ca^{2+} would fail to prevent the trifluoperazine-induced inhibition of G-cyclase activation. Furthermore, it has been reported that the rate of inhibition of Ca^{2+}/CaM-stimulated enzyme activity by the drug was increased by reducing concentrations of CaM, and that the apparent affinity of the CaM for G-cyclase decreased with increasing concentrations of the drug. The drug had little or no effect on the activities of Ca^{2+}-independent G-cyclase and *Tetrahymena* A-cyclase, which is not influenced by the Ca^{2+}/CaM system. These results indicate that trifluoperazine is a competitive inhibitor of *Tetrahymena* CaM, as observed with other CaM-dependent enzyme systems in various tissues (Weiss et al. 1980). In this respect, Shimizu et al. (1984) studied the interaction of trifluoperazine with *Tetrahymena* CaM using ^{19}F NMR. Their results suggested that, although the drug was located near a hydrophilic region of the CaM molecule in the absence of Ca^{2+}, under excess Ca^{2+} it was translocated to the neighborhood of a hydrophobic region of the CaM.

Kameyama et al. (1982) investigated the modification of G-cyclase activity in phospholipase A_2-treated membranes from *Tetrahymena*. The enzyme activity progressively decreased as the content of lysophosphatidylethanolamine increased due to phospholipase A_2-treatment, indicating that *Tetrahymena* G-cyclase requires an appropriate membrane phospholipid environment for the enzyme to function properly. This finding was consistent with earlier evidence that the G-cyclase activity in mammalian cells was enhanced by lysophosphatidylcholine but not by lysophosphatidylethanolamine (Shier et al. 1976).

With great effort, particulate G-cyclase was purified in some cases, including from rat liver (Haguenauer-Tsapis et al. 1981), bovine retinal rod outer segments (Koch 1991), and sea urchim sperm (Radany et al. 1983). Nagao and Nozawa (1987) partially purified the particulate G-cyclase in about a 22% yield from *Tetrahymena* plasma membranes using a digitonin-solubilizing method followed by DE52 column and CaM-Sepharose affinity chromatographies. Their findings of divalent cation dependency and kinetics indicated that the solubilized enzyme resembled the enzyme in the native membrane-bound form, and that a Ca^{2+}-dependent CaM-binding site was present on the solubilized enzyme. Klumpp et al. (1984) also partially purified *Paramecium* particulate G-cyclase.

A number of different G-cyclase receptors have been isolated and characterized, but the details of the mechanisms by which ligand binding stimulates and/or regulates G-cyclase activation need to be further clarified.

5 Structure and Intracellular Distribution of Calmodulin

Of some Ca^{2+}-binding proteins, only CaM has a broad distribution within the cell and throughout different tissues and species. Moreover, CaM alone is a multifunctional regulatory protein which activates, in a Ca^{2+}-dependent manner, a variety of the enzymes involved in many physiological processes of most, if not all, organisms. CaM, a heat- and acid-stable protein, was discovered as an activator of cyclic nucleotide PDE (Cheung 1970; Kakiuchi et al. 1970) and was identified as a Ca^{2+}-binding protein (Teo and Wang 1973). Thereafter, *Tetrahymena* CaM was purified to homogeneity using various methods and Ca^{2+}-dependently produced a great stimulation of mammalian brain cyclic nucleotide PDEs (Suzuki et al. 1979; Jamieson et al. 1979; Kumagai et al. 1980; Kakiuchi et al. 1981).

Tetrahymena CaM has many of the general features previously found in higher organisms, including a low molecular weight of ≈15 kDa (Suzuki et al. 1979; Jamieson et al. 1979; Kumagai et al. 1980), a low pI of 4.0 (Suzuki et al. 1979), heat- and acetone tolerance (Suzuki et al. 1979), a different electrophoretic mobility in Ca^{2+} (Suzuki et al. 1979), formation of complexes with troponin I in the presence of Ca^{2+}, and binding of phenothiazines (Nagao et al. 1981a). Blum et al. (1980) reported that *Tetrahymena* CaM mediated the Ca^{2+}-regulation of its ciliary dynein ATPase. The amino acid composition of *Tetrahymena* CaM is remarkably analogous to those of animal CaMs (Watterson et al. 1980), which contain an unusual amino acid trimethyllysine as well as high phenylalanine and low tyrosine contents, but not tryptophan and cystein.

Figure 7 illustrates the complete amino acid sequence of *Tetrahymena* CaM (Yazawa et al. 1981), compared with those of bovine brain CaM (Watterson et al. 1980) and *Paramecium* CaM (Schaefer et al. 1987). These CaMs possess four helix-loop-helix Ca^{2+}-binding domain structures. The loops consist of 12

```
 1                  10                  20         I      30
Ac-A D Q L T E E Q I A E F K E A F S L F D K D G N G T I T T K E L G T V M
Ac-A Q E L T E E Q I A E F K"E A F A L F D K D G D G T I T T K E L G T V M
Ac-A D Q L T E E Q I A E F K E A F S L F D K D G N G T I T T K E L G T V M

       40                  50                  60 II              70
     R S L G Q N P T E A E L Q D M I N E V D A D G N G T I D F P E F L S L M A
     R S L G Q N P T E A E L Q D M I N E V D A D G N G T I D F P E F L S L M A
     R S L G Q N P T E A E L Q D M I N E V D A D G N G T I D F P E F L T M M A

             80                  90                  III 100
     R K M K D T D S E E E L I E A F K V F D R D G N G L I T A A E L R H V M
     R K M K E Q D S E E E L I E A F K V F D R D G N G L I S A A E L R H V M
     R K M K D T D S E E E I R E A F R V F D K D G N G Y I S A A E L R H V M

    110                  120                  130        IV        140             148
     T N L G E K'L T D E E V D E M I R E A N I D G D G H I N Y E E F V R M M - A K-COOH
     T N L G E K'L T D D E V D E M I R E A D I D G D G H I N Y E E F V R M M V S K-COOH
     T N L G E K'L T D E E V D E M I R E A N I D G D G E V N Y E E F V Q M M T A K-COOH
```

Fig. 7. Amino acid sequences of *Tetrahymena*, *Paramecium*, and bovine brain calmodulins. Each has four helix-loop-helix Ca^{2+}-binding domains (*I, II, III* and *IV*), which are indicated by the *solid box*. *Upper panel Tetrahymena* calmodulin; *middle lane Paramecium* calmodulin; *lower lane* bovine brain calmodulin. Protozoan residues distinguishable from bovine protein are indicated by the *dotted box*. Sequence differences between the two protozoan calmodulins are shown by the *broken box*; *K'* trimethyllysine; *K"* dimethyllysine. One deletion at position 146 of *Tetrahymena* calmodulin is shown as a dash (–). (*Tetrahymena* calmodulin, Yazawa et al. 1981; *Paramecium* calmodulin, Schaefer et al. 1987; bovine brain calmodulin, Watterson et al. 1980)

amino acid residues, 6 of which have oxygen-containing side chains assumed to contribute oxygen ligands to Ca^{2+} at the vertices of an octahedron (Kretsinger 1976). *Tetrahymena* CaM was found to be composed of 147 amino acids and an acetylated amino-terminal. Furthermore, *Tetrahymena* CaM is mutually inconsistent with bovine brain CaM at 12 loci of amino acid compositions containing deletion of amino acid residue 147 near the carboxyl-terminal. The difference in amino acid sequences between the two protozoan and bovine CaMs appears to be large, compared with only three replacements between the amino acid residues of bovine brain and sea anemone CaMs (Takagi et al. 1980). Among amino acid replacements of the residues 70, 71, 85, 86, 90, 94, 101, 135, 136, 143, and 146, those at positions 86 (Arg→lle), 135 (Gln→His), and 143 (Gln→Arg) are considered to result in functional differences in the protein; e.g., a possible decrease in the affinity for Ca^{2+} of the Ca^{2+}-binding domain IV due to the introduction of a partial positive charge of His instead of γ-carbonyl oxygen of Gln 135, which is the predicted -Y octahedral binding coordinate in the domain IV. In bovine brain CaM, the amino acid residues 71–77 have been proposed for the site of interaction with PDE, resulting from its disability to activate PDE after the oxidation of methionine residues 71, 72, 76, and possibly 109 (Walsh and Stevens 1978). In *Tetrahymena* CaM, however, Met 71 was replaced by Leu, suggesting no relationship between the Met 71 and PDE activation. Kretsinger and Barry (1975) demonstrated that a hinge region (amino acid residues 75–80) between the Ca^{2+}-binding domains II and III as well as both the carboxyl-and amino-terminals was exposed to the surface area and would represent the site for Ca^{2+}-dependent interaction with enzymes.

To evaluate the functional significance of the carboxyl-terminal in G-cyclase activation, Nagao et al. (1990) engineered three types of CaMs which had point-mutations near the carboxyl-terminal, using cassette mutagenesis of rat brain CaM cDNA: Gln-143→Arg (CaM-A), Thr-146→deletion (CaM-D), and Gln-143→Arg/Thr-146→deletion (CaM-AD). Although these CaMs containing recombinant wild-type CaM (wCaM) activated rat brain cAMP PDE, only two mutated CaMs of CaM-A and CaM-AD with the Gln-143 replacement were Ca^{2+}-dependently able to stimulate *Tetrahymena* G-cyclase. These results and data on its inhibitory effects by a CaM antagonist W-7 suggest that Arg-143 is in a molecular region playing a pivotal role in the activation of *Tetrahymena* plasma membrane G-cyclase.

Nagao et al. (1981b) determined the subcellular distribution in *Tetrahymena* CaM (Table 4), and found that nearly all (93%) of the total amount of CaM was recovered in two soluble compartments, such as the ciliary and post-microsomal supernatant fractions, which are considered to be two major CaM reservoirs, with sevenfold higher specific activity in the former than in the latter. There is evidence that these two pools are metabolically isolated from each other to some extent. In addition, particulate CaM-binding activity was demonstrated to be accumulated largely in microsomes and to some extent in cilia and pellicles, with the highest specific activity (about 57%) in the former

Table 4. Subcellular distribution of particulate CaM-binding activity and CaM in *Tetrahymena*. (Nagao et al. 1981b)

	Protein		CaM-binding activity			CaM		
	mg	% of total	µg[^{3}H] CaM/ mg protein	µg[^{3}H] CaM	% of total	µg CaM/ mg protein	µg CaM	% of total
Cilia	3.8	0.7	0.45	1.7	1.8	1.3	5.1	1.0
Ciliary supernatant	31.8	5.5	–	–	–	6.7	212	43
Pellicles	138	24	0.12	16.6	18	0.1	13.4	2.7
Mitochondria	63.2	11	0.04	2.5	2.7	0.2	10.8	2.2
Microsomes	89.7	16	0.80	71.8	77.5	0.07	6.2	1.3
Postmicrosomal supernatant	247	43	–	–	–	1.0	243	50
Total calculated value	573.5	100.2		92.6	100	0.86[a]	490.5	100.2

[a]Total amount of CaM/total amount of protein.

(Table 4). The CaM and binding activities detected in the cilia and/or pellicles may be attributable to dynein ATPase and G-cyclase, respectively, because Ca^{2+}-dependent activation of these enzymes due to *Tetrahymena* CaM has been reported (Jamieson et al. 1979; Kakiuchi et al. 1981). Actually, G-cyclase is solubilized from *Tetrahymena* plasma membranes (Nagao and Nozawa 1987). The binding activity involved in microsomes remains unexplained. When *Tetrahymena* microsomes solubilized with Triton X-100 were eluated on a CaM-Sepharose chromatography, several proteins were found to interact with CaM in a Ca^{2+}-dependent manner (Nagao and Nozawa 1985), but they have not been identified yet. To date, the biological functions of the CaM-binding proteins in *Tetrahymena* microsomes are unknown. In animal cells, CaM-activated protein kinase has been reported in microsomal preparations (Famulski and Carafoli 1984), including sarcoplasmic reticulum (Campbell and MacLennan 1982). However, CaM-dependent protein kinase activity has not been detected in *Tetrahymena* microsomes. The *Tetrahymena* microsomal Ca^{2+}-pumping ATPase reported by Muto and Nozawa (1984) was also not stimulated by adding CaM. It is conceivable that the binding protein neither is a protein kinase nor plays a role in Ca^{2+} transport. In any case, it is expected that the major function of the CaM in *Tetrahymena* in relation to Ca^{2+} will be clarified by further work on this microsomal binding site.

Satir et al. (1980) recognized an immunocytochemical localization of the CaM in a cellular component corresponding to basal bodies at the point where the ciliary central microtubules terminate at the level of the cell surface in many protozoa. However, the subcellular fraction of the basal body is not clearly assigned. Using an indirect immunofluorescent method using anti-*Tetrahymena* CaM antibody, Suzuki et al. (1982) demonstrated that CaM was localized in the oral apparatus, cilia, basal bodies, the anterior end of the cell, and the contractile vacuole pores, and suggested the involvement of CaM in food vacuole formation accompanied by nutrient uptake, excretion of contractile vacuole contents associated with regulation of osmotic pressure, and in ciliary movement. This conclusion was supported by the remarkable suppression of these phenomena due to trifluoperazine treatment.

In *Tetrahymena*, some attempts have been made to detect Ca^{2+}-binding proteins other than CaM. Ohnishi and Watanabe (1983) found that *Tetrahymena* cilia contained both CaM and another type of Ca^{2+}-binding protein of 10 kDa (*Tetrahymena* Ca^{2+}-binding protein of 10 kDa, TCBP-10), and suggested that the first CaM family protein TCBP-10 and CaM play pivotal cooperative roles in the Ca^{2+}-regulation of ciliary movement in *Tetrahymena*. Recently, the third CaM family protein (TCBP of 23 kDa, TCBP-23) with Ca^{2+}-binding property has been cloned by cDNA analysis (Takemasa et al. 1990), and is expected to provide crucial clues for determining the mechanisms of Ca^{2+}-dependent phenomena. *Tetrahymena* CaM is almost as effective as mammalian CaM in activating CaM-related enzymes, such as rat brain A-cyclase, chicken gizzard myosin light chain kinase, erythrocyte Ca^{2+}/Mg^{2+}-ATPase, and plant NAD kinase (Kudo et al. 1982b). Concurrently, it is unique in its capability to activate G-cyclase; with the exception of the *Paramecium* protein, no other CaMs have activated the enzyme. Structural and immunochemical analyses of *Tetrahymena* CaM accompanied by additional use of the site-specific mutagenesis technique have the potential of providing further insight into how the structure of CaM is related to in vivo function and how Ca^{2+}/CaM modulation of cellular signal transduction might be coupled at a molecular level.

6 Conclusion

In this chapter we described the activation mechanisms of A- and G-cyclases closely associated with cyclic nucleotide metabolism and functional structures of the regulator protein CaM in *Tetrahymena* compared with a genus related *Paramecium*. Regarding the regulatory aspects of cellular functions, a variety of questions remain to be answered: (1) What are the functions of the unidentified CaM-binding proteins? (2) Is CaM an intracellular Ca^{2+} regulator? (3) What is the molecular interrelation between the Ca^{2+}/CaM system, the cyclic nucleotides, and both cyclases? (4) What is the functional involvement of these factors in a number of types of physical and biological behavior in eukaryotic cells? *Tetrahymena* is a valuable model system which is ripe for use in the intensive study of membrane-mediated cellular regulation, including analysis of the morphological, biochemical, biophysical, and genetic aspects of various biological phenomena in eukaryotic cells. It is expected that, using this organism, researchers will gain a better understanding of the biochemistry and physiology of the cAMP and cGMP signal transduction systems.

References

Aomine M (1976) Further studies on the mechanism of uptake of D-glucose by *Tetrahymena pyriformis* GL. Comp Biochem Physiol A 55: 159–163

Blum JJ, Hayes A, Jamieson GA, Vanaman TC (1980) Calmodulin confers calcium sensitivity on ciliary dynein ATPase. J Cell Biol 87: 386–397

Böhme E, Jung R, Mechler I (1974) Guanylate cyclase in human platelets. Methods Enzymol 38: 199–202

Campbell KS, MacLennan DH (1982) A calmodulin-dependent protein kinase system from skeletal muscle sarcoplasmic reticulum: phosphorylation of a 60,000-dalton protein. J Biol Chem 257: 1238–1246

Cheung WY (1970) Cyclic 3′,5′-nucleotide phosphodiesterase. Demonstration of an activator. Biochem Biophys Res Commun 38: 533–538

Coffey RG, Hadden EM, Lopez C, Hadden JW (1978) cGMP and calcium in the initiation of cellular proliferation. Adv Cyclic Nucleotide Res 9: 661–676

Dickinson JR, Graves MG, Swoboda BEP (1976) Cyclic AMP metabolism in the cell cycle of *Tetrahymena pyriformis*. FEBS Lett 65: 152–154

Dickinson JR, Graves MG, Swoboda BEP (1977) Induction of division synchrony in *Tetrahymena pyriformis* by a single hypoxic shock. Its use in elucidating control of the cell cycle by adenosine 3′: 5′-monophosphate. Eur J Biochem 78: 83–87

Eckert R (1972) Bioelectric control of ciliary activity. Science 176: 473–481

Famulski KS, Carafoli E (1984) Calmodulin-dependent protein phosphorylation and calcium uptake in rat-liver microsomes. Eur J Biochem 141: 15–20

Garbers DL, Gray JP (1974) Guanylate cyclase from sperm of the sea urchin, *Strongylocentrotus purpuratus*. Methods Enzymol 38: 196–199

Goldberg ND, O'Dea RF, Haddox MK (1973) Cyclic GMP. Adv Cyclic Nucleotide Res 3: 155–223

Gray NCC, Dickinson JR, Swoboda BEP (1977) Cyclic GMP metabolism in *Tetrahymena pyriformis* synchronized by a single hypoxic shock. FEBS Lett 81: 311–314

Gunderson RE, Thompson GA Jr (1983) Factors influencing the pattern of dopamine secretion in *Tetrahymena pyriformis*. Biochim Biophys Acta 755: 186–194

Haguenauer-Tsapis R, Ben Salah A, Lacombe M-L, Hanoune J (1981) Trypsin solubilization of rat liver membrane-bound guanylate cyclase results in a form kinetically distinct from the cytosolic enzyme. J Biol Chem 256: 1651–1655

Jamieson GA, Vanaman TC, Blum JJ (1979) Presence of calmodulin in *Tetrahymena*. Proc Natl Acad Sci USA 76: 6471–6475

Janakidevi K, Dewey VC, Kidder GW (1966) The biosynthesis of catecholamines in two genera of protozoa. J Biol Chem 241: 2576–2578

Kakiuchi S, Yamazaki R, Nakajima H (1970) Properties of a heat-stable phosphodiesterase activating factor isolated from brain extracts. Studies on cyclic 3′,5′-nucleotide phosphodiesterase II. Proc Jpn Acad 46: 589–592

Kakiuchi S, Sobue K, Yamazaki R, Nagao S, Umeki S, Nozawa Y, Yazawa M, Yagi K (1981) Ca^{2+}-dependent modulator proteins from *Tetrahymena pyriformis*, sea anemone, and scallop and guanylate cyclase activation. J Biol Chem 256: 19–22

Kameyama Y, Kudo S, Ohki K, Nozawa Y (1982) Differential inhibitory effects by phospholipase A_2 on guanylate and adenylate cyclases of *Tetrahymena* plasma membranes. Jpn J Exp Med 52: 183–192

Kariya K, Saito K, Iwata H (1974) Adrenergic mechanism in *Tetrahymena* III. Cyclic adenosine 3′,5′-monophosphate and cell proliferation. Jpn J Pharmacol 24: 129–134

Kassis S, Kindler SH (1975) Dispersion of epinephrine sensitive and insensitive adenylate cyclase from the ciliate *Tetrahymena pyriformis*. Biochim Biophys Acta 391: 513–516

Kassis S, Zeuthen E (1979) Adenylate cyclase and cyclic AMP through the cell cycle of *Tetrahymena*. Exp Cell Res 124: 73–78

Klumpp S, Schultz JE (1982) Characterization of a Ca^{2+}-dependent guanylate cyclase in the excitable ciliary membrane from *Paramecium*. Eur J Biochem 124: 317–324

Klumpp S, Gierlich D, Schultz JE (1984) Adenylate cyclase and guanylate cyclase in the excitable ciliary membrane from *Paramecium*: separation and regulation. FEBS Lett 171:95–99

Koch K-W (1991) Purification and identification of photoreceptor guanylyl cyclase. J Biol Chem 266: 8634–8637

Kohidai L, Barsony J, Roth J, Marx SJ (1992) Rapid effects of insulin on cyclic GMP location in an intact protozoan. Experientia 48: 476–481

Kretsinger RH (1976) Calcium binding proteins. Annu Rev Biochem 45: 239–266

Kretsinger RH, Barry CD (1975) The predicted structure of the calcium binding component of troponin. Biochim Biophys Acta 405: 40–52

Krupinski J, Coussen F, Bakalyar HA, Tang W-J, Feinstein PG, Orth K, Slaughter C, Reed RR, Gilman AG (1989) Adenylyl cyclase amino acid sequence: possible channel- or transporter-like structure. Science 244: 1558–1564

Kudo S, Nakazawa K, Nozawa Y (1980) Studies on cyclic nucleotide metabolism in *Tetrahymena pyriformis*: partial characterization of cyclic AMP- and cyclic GMP-dependent phosphodiesterase. J Protozool 27: 342–345

Kudo S, Nagao S, Kameyama Y, Nozawa Y (1981a) Growth-associated changes in cyclic nucleotide enzymes in *Tetrahymena*: involvement of calmodulin. Cell Differ 10: 237–242

Kudo S, Nagao S, Kasai R, Nozawa Y (1981b) Cell cycle-associated changes of guanylate cyclase activity in synchronized *Tetrahymena*: a possible involvement of calmodulin in its regulation. J Protozool 28: 165–167

Kudo S, Ohnishi K, Muto Y, Watanabe Y, Nozawa Y (1981c) *Paramecium* calmodulin can stimulate membrane-bound guanylate cyclase in *Tetrahymena*. Biochem Int 3: 255–263

Kudo S, Nakazawa K, Nagao S, Nozawa Y (1982a) Calmodulin alters the cation requirement of membrane-bound guanylate cyclase in *Tetrahymena*. Jpn J Exp Med 52: 193–200

Kudo S, Muto Y, Nagao S, Naka M, Hidaka H, Sano M, Nozawa Y (1982b) Specificity of *Tetrahymena* calmodulin in activation of calmodulin-regulated enzyme. FEBS Lett 149: 271–276

Kudo S, Muto Y, Nozawa Y (1985) Regulation by calcium of hormone-insensitive adenylate cyclase and calmodulin-dependent guanylate cyclase in *Tetrahymena* plasma membrane. Comp Biochem Physiol 80B: 813–816

Kumagai H, Nishida E, Ishiguro K, Murofushi H (1980) Isolation of calmodulin from the protozoan, *Tetrahymena pyriformis*, by the use of a tubulin-Sepharose 4B affinity column. J Biochem 87: 667–670

Kuno T, Yoshida N, Tanaka C, Kasai R, Nozawa Y (1981) Immunocytochemical localization of cyclic AMP in *Tetrahymena*. Experientia 37: 411–413

Lovely JR, Threlfall RJ (1979) The activity of guanylate cyclase and cyclic GMP phosphodiesterase during synchronous growth of the cellular slime mould *Physarum polycephalum*. Biochem Biophys Res Commun 86: 365–370

May DC, Ross EM, Gilman AG, Smigel MD (1985) Reconstitution of catecholamine-stimulated adenylate cyclase activity using three purified proteins. J Biol Chem 260: 15829–15833

Murofushi H (1974) Protein kinases in *Tetrahymena* cilia. II. Partial purification and characterization of adenosine 3′,5′-monophosphate-dependent and guanosine 3′,5′-monophosphate-dependent protein kinases. Biochim Biophys Acta 370: 130–139

Muto Y, Nozawa Y (1984) Biochemical characterization of ($Ca^{2+}+Mg^{2+}$)-ATPase in *Tetrahymena* microsomes. Biochim Biophys Acta 777: 67–74

Nagao S, Nozawa Y (1985) Calmodulin-binding proteins of *Tetrahymena* microsomal membranes. Comp Biochem Physiol 82B: 689–693

Nagao S, Nozawa Y (1987) Properties of digitonin-solubilized calmodulin-dependent guanylate cyclase from the plasma membranes of *Tetrahymena pyriformis* NT-1 cells. Arch Biochem Biophys 252: 179–187

Nagao S, Suzuki Y, Watanabe Y, Nozawa Y (1979) Activation by a calcium-binding protein of guanylate cyclase in *Tetrahymena pyriformis*. Biochem Biophys Res Commun 90: 261–268

Nagao S, Kudo S, Nozawa Y (1981a) Effects of phenothiazines on the membrane-bound guanylate cyclase and adenylate cyclase in *Tetrahymena pyriformis*. Biochem Pharmacol 30: 2709–2712

Nagao S, Banno Y, Nozawa Y, Sobue K, Yamazaki R, Kakiuchi S (1981b) Subcellular distribution of calmodulin and calmodulin-binding sites in *Tetrahymena pyriformis*. J Biochem 90: 897–899

Nagao S, Matsuki S, Kanoh H, Ozawa T, Yamada K, Nozawa Y (1990) Site-directed mutagenesis of glutamine residue of calmodulin. Activation of guanylate cyclase of *Tetrahymena* plasma membrane. J Biol Chem 265: 5926–5929

Nakazawa K, Shimonaka H, Nagao S, Kudo S, Nozawa Y (1979) Magnesium-sensitive guanylate cyclase and its endogenous activating factor in *Tetrahymena pyriformis*. J Biochem 86: 321–324

Nandini-Kishore SG, Thompson GA Jr (1979) Increased levels of adenosine 3′,5′-cyclic monophosphate in *Tetrahymena* stimulated by glucose and mediated by Ca^{2+} and epinephrine. Proc Natl Acad Sci USA 76: 2708–2711

Nozawa Y, Thompson GA Jr (1971) Studies of membrane formation in *Tetrahymena pyriformis*. II. Isolation and lipid analysis of cell fractions. J Cell Biol 49: 712–721

Nozawa Y, Umeki S (1988) Regulation of membrane fluidity in unicellular organisms. In: Aloia RC, Curtain CC, Gordon LM (eds) Physiological regulation of membrane fluidity. Alan R Liss, New York, pp 239–257

O'Day DH (1981) Modes of cellular communication and sexual interactions in eukaryotic microbes. In: O'Day DH, Horgen PA (eds) Sexual interactions in eukaryotic microbes. Academic Press, New York, pp 3–17

Ohnishi K, Watanabe Y (1983) Purification and some properties of a new Ca^{2+}-binding protein (TCBP-10) present in *Tetrahymena* cilium. J Biol Chem 258: 13978–13985

Radany EW, Gerzer R, Garbers DL (1983) Purification and characterization of particulate guanylate cyclase from sea urchin spermatozoa. J Biol Chem 258: 8346–8351

Ross EM (1982) Phosphatidylcholine-promoted interaction of the catalytic and regulatory proteins of adenylate cyclase. J Biol Chem 257: 10751–10758

Rozensweig Z, Kindler SH (1972) Epinephrine and serotonin activation of adenyl cyclase from *Tetrahymena pyriformis*. FEBS Lett 25:221–223

Satir BH, Garofalo RS, Gilligan DM, Maihle NJ (1980) Possible functions of calmodulin in protozoa. Ann N Y Acad Sci 356: 83–91

Schaefer WH, Lukas TJ, Blair IA, Schultz JE, Watterson DM (1987) Amino acid sequence of a novel calmodulin from *Paramecium tetraurelia* that contains dimethyllysine in the first domain. J Biol Chem 262: 1025–1029

Schultz JE, Klumpp S (1983) Adenylate cyclase in cilia from *Paramecium*: localization and partial characterization. FEBS Lett 154: 347–350

Schultz JE, Schönborn C (1994) Cyclic AMP formation in *Tetrahymena pyriformis* is controlled by a K^{+}-conductance. FEBS Lett 356: 322–326

Schultz JE, Schönefeld U, Klumpp S (1983) Calcium/calmodulin-regulated guanylate cyclase and calcium-permeability in the ciliary membrane from *Tetrahymena*. Eur J Biochem 137: 89–94

Schultz JE, Pohl T, Klumpp S (1986) Voltage-gated Ca^{2+} entry into *Paramecium* linked to intraciliary increase in cyclic GMP. Nature 322: 271–273

Schultz JE, Uhl DG, Klumpp S (1987) Ionic regulation of adenylate cyclase from the cilia *Paramecium tetraurelia*. Biochem J 246: 187–192

Schultz JE, Klumpp S, Benz R, Schürhoff-Goeters WJ Ch, Schmid A (1992) Regulation of adenylyl cyclase from *Paramecium* by an intrinsic pottasium conductance. Science 255: 600–603

Seifert W, Rudland PS (1974) Possible involvement of cyclic GMP in growth control of cultured mouse cells. Nature 248: 138–140

Shier WT, Baldwin JH, Nilsen-Hamilton M, Hamilton RT, Thanassi NM (1976) Regulation of guanylate and adenylate cyclase activities by lysolecithin. Proc Natl Acad Sci USA 73: 1586–1590

Shimizu T, Hatano M, Muto Y, Nozawa Y (1984) Interaction of trifluoperazine with *Tetrahymena* calmodulin. A ^{19}F NMR study. FEBS Lett 166: 373–377

Shimonaka H, Nozawa Y (1977) Subcellular distribution and thermally-induced transition of adenylate cyclase activity in thermotolerant *Tetrahymena* surface membranes. Cell Struct Funct 2: 81–89

Suzuki Y, Hirabayashi T, Watanabe Y (1979) Isolation and electrophoretic properties of a calcium-binding protein from the ciliate *Tetrahymena pyriformis*. Biochem Biophys Res Commun 90: 253–260

Suzuki Y, Ohnishi K, Hirabayashi T, Watanabe Y (1982) *Tetrahymena* calmodulin. Characterization of an anti-*Tetrahymena* calmodulin and the immunofluorescent localization in *Tetrahymena*. Exp Cell Res 137: 1–14

Takagi T, Nemoto T, Konishi K, Yazawa M, Yagi K (1980) The amino acid sequence of the calmodulin obtained from sea anemone (*Metridium seline*) muscle. Biochem Biophys Res Commun 96: 377–381

Takemasa T, Takagi T, Kobayashi T, Konishi K, Watanabe Y (1990) The third calmodulin family protein in *Tetrahymena*. Cloning of the cDNA for *Tetrahymena* calcium-binding protein of 23 kDa (TCBP-23). J Biol Chem 265: 2514–2517

Teo TS, Wang JH (1973) Mechanism of activation of a cyclic adenosine 3′:5′-monophosphate phosphodiesterase from bovine heart by calcium ions. J Biol Chem 248: 5950–5955

Umeki S, Nozawa Y (1984) Repression by dexamethasone of epinephrine-induced modulation of the fatty acyl-CoA desaturase system in *Tetrahymena* microsomes. Eur J Biochem 142: 355–359

Umeki S, Nozawa Y (1985) A possible cyclic AMP-mediated regulation of microsomal fatty acyl-CoA desaturation system in *Tetrahymena* microsomes. Biochim Biophys Acta 835: 514–526

Voichick J, Elson C, Granner D, Shrago E (1973) Relationship of adenosine 3′,5′-monophosphate to growth and metabolism of *Tetrahymena pyriformis*. J Bacteriol 115: 68–72

Walsh M, Stevens FC (1978) Chemical modification studies on the Ca^{+2}-dependent protein modulator. The role of methionine residues in activation of cyclic nucleotide phosphodiesterase. Biochemistry 17: 3924–3928

Watanabe Y, Nozawa Y (1982) Possible roles of calmodulin in a ciliated protozoan *Tetrahymena*. In: Cheung WY (ed) Calcium and cell function, vol II. Academic Press, New York, pp 297–323

Watterson DM, Sharief FS, Vanaman TC (1980) The complete amino acid sequence of the Ca^{2+}-dependent modulator protein (Calmodulin) of bovine brain. J Biol Chem 255: 962–975

Weiss B, Levin RM (1978) Mechanism for selectively inhibiting the activation of cyclic nucleotide phosphodiesterase and adenylate cyclase by antipsychotic agents. Adv Cyclic Nucleotide Res 9: 285–303

Weiss B, Prozialeck W, Cimino M (1980) Acute and chronic effects of psychoactive drugs on adrenergic receptors and calmodulin. Adv Cyclic Nucleotide Res 12: 213–225

Wong, Garbers (1992) Receptor guanylyl cyclases. J Clin Invest 90: 299–305

Yazawa M, Yagi K, Toda H, Kondo K, Narita K, Yamazaki R, Sobue K, Kakiuchi S, Nagao S, Nozawa Y (1981) The amino acid sequence of the *Tetrahymena* calmodulin which specifically interacts with guanylate cyclase. Biochem Biophys Res Commun 99: 1051–1057

Zimmerman S, Zimmerman AM, Laurence H (1981) Effect of Δ^9-tetrahydrocannabinol on cyclic nucleotides in synchronously dividing *Tetrahymena*. Can J Biochem 59: 489–493

Signal Peptide-Induced Sensory Behavior in Free Ciliates: Bioassays and Cellular Mechanisms

V. Leick, M. Grave, and P. Hellung-Larsen[1]

1 Introduction

Cellular chemosensory and other kinds of behavior of protozoa can be considered a fundamental aspect of cellular activity equivalent to cellular growth and sexual reproduction. For motile protozoan cells, the type of chemicals which act as attractants and repellents in natural environments probably reflects the ecological niches which a particular organism selects as its favorite environment, in particular when it has to coexist with other organisms and still survive and maintain its identity. The different chemical stimuli signify the presence of food, mates, toxic conditions, hosts, etc.

We have particularly studied the effect of signal peptides on the chemosensory phenotype of the ciliated protozoan *Tetrahymena,* a phagotrophic ciliate, which has a long evolutionary history and, according to David Nanney, "may have been fairly lonely denizens of the lakes and seas some 2 billion years ago when they radiated into the ecosystem and exploited their basic eukaryotic equipment and their, for the time, enormous size" (Nanney 1982). The present chapter is an attempt to review our work and to relate it to similar work in other ciliated free-living protozoa.

Of the more than 7000 species of ciliate protozoans (Corliss 1979), *Tetrahymena* is one of the most readily obtainable in axenic culture. Since the first report by Lwoff (Lwoff 1923) on his success in growing *Tetrahymena pyriformis* in bacteria-free axenic broth culture, many strains of this organism have been studied. They all require 11 amino acids in their diet, a purine and a pyrimidine source, and various vitamins and salts (Holz 1973). However, the fastest and most reproducible growth rates are obtained with complex broth media. These media are also efficient chemoattractants (Leick and Helle 1983; Leick and Hellung-Larsen 1985). It is the amino acid fraction of defined media and peptide fraction of the complex growth media which are the major chemoattractants for *Tetrahymena* (Almagor et al. 1981; Levandowsky et al. 1984; Leick and Hellung-Larsen 1985).

[1]Institute of Medical Biochemistry and Genetics, Department of Biochemistry B, The University of Copenhagen, The Panum Institute, Blegdamsvej 3C, 2200 Copenhagen, Denmark

Growth of other ciliates like *Paramecium* is also, although less readily, obtained in crude axenic media (Soldo and Van Wagtendonk 1969). *Paramecium* detects folate, lactate, acetate, and cyclic AMP as chemoattractants, and these compounds probably indicate the presence of food (Van Houten 1978, 1992). Cytochrome c was reported to act as a chemorepellent in *Paramecium* (Francis and Hennessey 1995). The amino acid glutamate was identified as an attractant, and the receptors for this compound identified on the cilia (Preston and Usherwood 1988).

The ciliate *Euplotes* uses mating pheromones in the mating process (Luporini and Miceli 1986; Van Houten 1994). In the marine species *E. raikovi* the pheromones are homodimers of 38–40 amino acid peptides which bind specifically to cell receptor sites (Ortenzi et al. 1990).

In the ciliate *Blepherisma japonicum*, cell accumulation related to the mating process occurs induced by the gamones (Miyake 1981). At low concentrations, the gamones (gamone I is a glycoprotein) serve as chemoattractants for cells of opposite mating types (Honda and Miyake 1975). In the slime molds *Dictoystelium discoideum* and *Polyspondylium violaceum*, chemoattraction and stalk development are controlled by two different chemoattractants (cAMP and the dipeptide glorin; Gerisch 1987; De Wit et al. 1988). *Entamoeba histolytica* migrates towards polysaccharides and other products that might simulate the host lumen where the trophozoites will establish themselves (Bailey et al. 1985).

The present chapter will emphasize and discuss the description of the technical and physiological aspects of assaying the chemosensory behavior of *Tetrahymena* by a survey of the different bioassays used to measure biological activity of the signal peptides. Moreover, the possible cellular mechanisms utilized by the cells to modify their swimming behavior during the chemosensory response will be formulated, as well as a reevaluation of the field in relation to a previous review (Leick and Hellung-Larsen 1992).

2 Peptide Signals in Ciliates

2.1 Signal Peptides Found Intracellularly

Vertebrate-like peptide hormone substances similar to insulin, somatostatin, ACTH, ß-endorphin, calcitonin (salmon type), relaxin, and arginine vasotocin have been found inside *Tetrahymena* by radioimmunological techniques (for review see, Roth et al. 1986; LeRoith et al. 1986). More recently, the presence of ß-endorphin-like immunoreactivity (about 10 pmol/g protein) was found using anti ß-endorphin-IgG-Sepharose (referred to as unpublished experiments in Renaud et al. 1991). The cells were grown in defined synthetic growth medium without peptides, so these immunoreacting substances could not have originated from the growth medium. Also, the partial characterization of the receptor protein for ß-endorphin in *Tetrahymena pyriformis* as a 110-kDa protein (with properties similar to the receptor in rat brain and the leech) has been carried out (O'Neill et al. 1988; Zipser et al. 1988). It has emphasized the potential usefulness of *Tetrahymena* for studies of the signal peptide/receptor physiology.

2.2 Signal Peptides Having Sensory Effects

Compared to the rather limited number of studies which have been carried out to characterize the biochemistry of peptide-induced signaling events in ciliates, much more information is available regarding the physiology, pharmacology, and cell biology of the sensory impact of peptides on the sensing system of *Tetrahymena*, in particular. A number of metazoan peptide hormones are chemoattractants as shown in Table 1, or have other physiological effects as compiled in Table 2.

2.2.1 Physiological Effects of Insulin on Tetrahymena

Since it was found that *Tetrahymena* is able to synthesize certain hormone-like substances, for example, the neurohormones epinepherine and norepinepherine (Janakidevi et al. 1966), and to respond to amino acid derivatives like the thyroid hormone triiodothyronine (Blum 1967), it was obvious to study if *Tetrahymena* is also influenced by peptide or protein hormones.

Several studies have provided physiological evidence for the ability of *Tetrahymena* to respond to hormonal concentrations of insulin (Tables 1 and 2). Csaba and Lantos (1975) studied the effect of insulin on the uptake of glucose in *Tetrahymena*. They found that the uptake of glucose by starved *T. pyriformis* was markedly increased if bovine insulin (350 nM) was added simultaneously. Later, it was reported that cells grown in a chemically defined medium contain compounds similar to insulin and other peptide hormones (for review see, LeRoith et al. 1986). Christopher and Sundermann (1992) found that *T. pyriformis* is able to bind insulin at its surface membrane. After fixation of cells from a culture grown in a broth medium, the cells were washed in PBS buffer and incubated with porcine insulin (8.7 μM) in PBS for 1 h. Fluorescence photomicroscopy revealed that insulin was bound to the ciliary surface and that the binding of FITC-insulin was reduced by prior incubation with unlabeled insulin. Koppelhus et al. (1994b) found that insulin in nanomolar concentrations is a significant chemoattractant for *Tetrahymena*. The two-phase assay (see Fig. 1) was carried out at 28 °C with *T. thermophila* cells starved for 24 h. In this assay insulin showed maximal effect at a concentration of 170 nM. The chemoattractant effect of insulin on *T. pyriformis* kept in broth medium was studied by Köhidai et al. (1994). They used the capillary assay (Leick and Helle 1983) to test if the cells were attracted by addition of 10^{-12}–10^{-6} M insulin. Maximal chemoattraction was observed at 0.1 nM insulin. The pharmacological evidence for the presence of receptors for insulin on the

Table 1. Chemosensory effects of peptides/proteins on *Tetrahymena*

Compound	MW	Concentration	Species	Effect	Assay	Reference[b]
α-Endorphin	1 746	nanomolar range	Therm.	Attr. (s)[a]	Capillary II	1
ß-Endorphin	3 465	7 nM	Therm.	Attr. (s)	Two-phase	1
Cytochrome c	12 500	160 nM	Pyr.	Attr. (s)	Two-phase	2
DansMLP	645	3 μM	Therm.	Attr. (s)	Capillary II	3
EGF	6 045	32 μM	Pyr.	Attr. (s)	Two-phase	2
FGF	13 000	400 nM	Pyr.	Attr. (s)	Field	4
FGF	13 000	75 nM	Therm.	Attr. (s)	Two-phase	1
fMLP	438	11 μM	Therm.	Attr. (s)	Capillary I	5
fMLP	438	11 μM	Therm.	Attr. (s)	Screening	5
fMLP	438	4.6 μM	Therm.	Attr. (s)	Capillary II	3
fMLP	438	2.5 μM	Pyr.	Attr. (s)	Two-phase	2
fMLPdansLys	799	2.5 μM	Therm.	Attr. (s)	Capillary II	3
fMLPLys	566	3.5 μM	Therm.	Attr. (s)	Capillary II	3
fMLPP	585	3.5 μM	Therm.	Attr. (s)	Capillary II	3
Hexokinase	102 000	50 nM	Pyr.	Attr. (s)	Two-phase	2
Histone 3	15 324	65 nM	Pyr.	Attr. (s)	Two-phase	2
Histone 4	11 282	45 nM	Pyr.	Attr. (s)	Two-phase	2
Insulin	5 750	170 nM	Therm.	Attr. (s)	Two-phase	1
Insulin	5 778	0.1 nM	Pyr.	Attr. (g)[c]	Capillary I	6
Lysozyme	14 400	70 nM	Pyr.	Attr. (s)	Two-phase	2
Met-enkephalin	574	200 μM	Therm.	Attr. (s)	Two-phase	1
Oxytocin	1 007	10 nM	Pyr.	Rep.[d]	Capillary I	7
PDGF	30 000	160 pM	Pyr.	Attr. (s)	Capillary I	8
PDGF	30 000	2.5 pM[e]	Therm.	Attr. (s)	Capillary I	6
PDGF	30 000	2.5 pM	Therm.	Attr. (s)	Screening	6
PDGF	30 000	160 pm	Therm. or pyr.	Attr. (s)	Filter paper	9
PDGF	30 000	3 nm	Therm.	Attr. (s)	Two-phase	1
Serum albumin	68 000	70 nM	Pyr.	Attr. (s)	Two-phase	2
Vasopressin	1 084	1 pM	Pyr.	Rep.O	Capillary I	7

[a]Chemoattraction of starved cells.
[b]1, Koppelhus et al. (1994b); 2, Hellung-Larsen et al. (1986); 3, Leick (1992); 4, Hellung-Larsen et al. (1990); 5, Leick and Hellung-Larsen (1985); 6, Köhidai et al. (1994); 7, Köhidai and Csaba (1995); 8, Andersen et al. (1984); 9, Leick et al. (1990).
[c]Chemoattraction of cells growing in broth medium (1% Bacto tryptone and 0.1% yeast extract).
[d]Chemorepulsion of cells growing in broth medium (1% tryptone and 0.1% yeast extract).
[e]By addition of 1.5% pure PDGF from Speywood Laboratories.

surface of *Tetrahymena* was further supported by the results of Kovács et al. (1989), who tested the effect of insulin on the level of guanylate cyclase activity in *T. pyriformis*. When cells from a culture grown in broth medium were incubated with 1 μM insulin for 1 h, the activity of guanylate cyclase increased by almost 40% compared to the unstimulated control. Köhidai et al. (1992) studied if addition of insulin changes the content and localization of cGMP in *T. pyriformis*. Addition of 10^{-12}–10^{-6} M insulin to a chemically defined medium caused accumulation of cGMP in the cells. Moreover, exposure to insulin

Table 2. Results from experiments where other biological effects of peptides/proteins have been studied on *Tetrahymena*

Compound	MW	Concentration	Species	Effect	Assay	References[i]
ACTH	4 567	300 nM	Therm.	IS[a]	Dilution[b]	1
BSA	69 000	15 nM	Therm.	IS	Dilution	1
DansMLP	645	4 nM	Therm.	I[c]	Microscopy	2
EGF	6 045	0.1 nM	Therm.	IS	Dilution	1
Insulin	5 750	170 nM	Therm.	IS	Dilution	1
NGF	130 000	0.01 nM	Therm.	IS	Dilution	1
Opiates		nanomolar range	Therm.	PI[d]	Microscopy	3
Oxytocin	1 007	1 μM	Pyr.	RCV[e]	Microscopy	4
PDGF	30 000	Unknown[f]	Pyr.	SN[g]	In vivo labeling[h]	5
PDGF	30 000	0.1 nM	Therm.	IS	Dilution	1

[a]IS: Increased survival of strongly diluted cultures in defined medium.
[b]After inoculation of 25 cells in 1 ml of the defined medium described in (Szablewski et al. 1991) (with or without glucose) the listed compounds prevented cell death in assays with a medium-air interphase. Addition of one of these compounds α-endorphin (140 nM), α-melanocyte stimulating hormone (60 nM), Arg-vasopressin (23 μM), Arg-vasotocin (24 μM), ß-endorphin (70 nM), bombesin (600 nM), fMLP (1.1 μM), glucagon (300 nM), leucine enkephalin (3.2 μM), physalaemin (800 nM), or suramin (1.5 μM) did not prevent total cell death.
[c]Internalization. Fluorescence microscopy showed that dansMLP was internalized preferentially into small vesicles. These vesicles were much smaller than food vacuoles.
[d]PI Phagocytosis inhibition.
[e]RCV reduces activity of contractile vacuole. The time interval between two systolic contractions of the contractile vacuole was increased by 74% upon primary exposure of the cells to this compound.
[f]The effect of partially purified PDGF was tested. The protein concentration was 50 ng/ml.
[g]SN stimulation of nucleic acid synthesis.
[h]Detection of acid-precipitable counts after incubation of starved cells with labeled DNA or RNA precursors.
[i]1, Kristiansen et al. (in prep.); 2, Leick (1992); 3, De Jesus and Renaud (1989); 4, Csaba and Kovács (1992); 5, Andersen et al. (1984).

caused a rapid time-dependent change in the subcellular localization of cGMP. The maximal effects were observed at 0.1 nM insulin. The cells responded to incubation with porcine, chicken, or human insulin or human proinsulin.

Insulin has effects on cellular growth and survival as well. When *Tetrahymena thermophila* cells are grown in a chemically defined medium (Szablewski et al. 1991) the size of the inoculum is critical for the survival of the cells (Christensen and Rasmussen 1992). When adding 1 μM bovine insulin to a low inoculum of *T. pyriformis* in defined medium, cell death and disintegration were prevented (Christensen 1993). Recently, it was found (Grave et al. 1995; Kristiansen et al., in prep.) that the cell death (and lysis) at low inocula is seen only in cultures with a liquid/air or liquid/oil surface. Addition of a number of peptide hormones or growth factors (insulin, NGF, EGF, PDGF, ACTH) or bovine serum albumin to such cultures rescued the cells

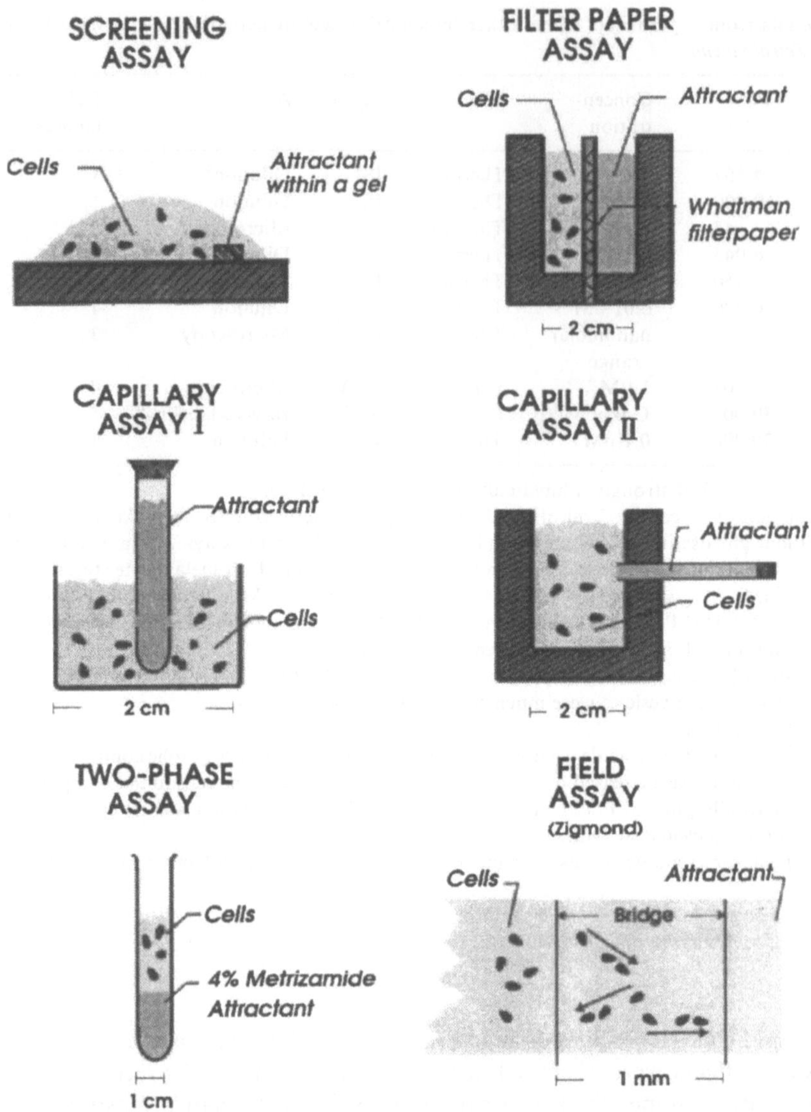

Fig. 1. Schematic representation of the assays used for qualitative and quantitative analysis of the chemosensory response of *Tetrahymena*. For details see text of the references given in Table 1

from the disintegration caused by the medium/air interphase (see Table 2). Moreover, addition of insulin (in the micromolar range) to cultures of *Tetrahymena thermophila* grown in this chemically defined medium also had a slight but significant stimulatory effect on cell growth (Hagemeister, unpubl. results). When the cells are starved in 10 mM HEPES buffer the addition of insulin (in the micromolar range) also resulted in a distinct reduction of the cell volume decrease (Hellung-Larsen, unpubl. results).

3 Bioassays Measuring Peptide-Induced Changes of Cell Behavior and Ciliary Activity

3.1 Population Assays for Chemoattraction

Several bioassays to measure the chemosensory response of cell suspensions of ciliates to signal peptides and other chemoattractants have been developed. Some of these methods which have been used in this laboratory are shown schematically in Fig. 1. In general, the cell population is brought in contact with a gradient of chemoattractant established by diffusion in various different technical setups as described below. The response of the cell population is then scored by measuring the number of cells reacting to the chemotactic gradient in relation to an unstimulated control.

In *the screening assay* the test substance is solidified in various ways by adding it to acrylamide, agarose, or gelatin before gelification. Cells (usually cells starved in 10 mM TRIS- or HEPES-buffer at pH 7.4) are brought in contact with the solidified attractant, and if the concentration of cells is > 50 000 cells/ml, the attraction or repulsion of the cells can be visualized in the stereomicroscope (or even by the naked eye) within 10–15 min at room temperature as a cloud of cells accumulating at the gel. The cells usually stay there for several hours until diffusion eliminates the gradient. The assay works well at room temperature, but the sensitivity is higher at 30–37 °C. The procedure is simple and rapid, but the method is qualitative and useful only for screening purposes. To choose a proper gel substance it must be considered that (1) polyacrylamide is toxic to the cells, (2) agarose forms a gel with a rather open structure, which leads to formation of a gradient with a short lifetime, (3) gelatin is easy to handle and forms a suitable tight gel structure, but some commercial products of gelatin are weak attractants by themselves. The extent of diffusion from the gels can be estimated by use of colored diffusable substances (crystal violet, hemoglobin, etc.). In short, the gel assay has mainly been used for screening of new substances. The *capillary assay* (version 1) is a capillary quantitative procedure where the cellular response is measured using an electronic cell counter (Leick and Helle 1983). Gradients of attractants or repellents are established between an inner compartment where the attractant is placed (usually a Plexiglas tube) and an outer well where (starved) cells are located. The tubes have capillary holes with a diameter of 0.5 mm and each tube has 16 holes (Leick and Helle 1983), but 8 holes may be sufficient (W. Pauli, pers. comm.). The accumulation of cells is usually allowed to occur for 2–3 h at room temperature. The cells become trapped in the inner compartment and can be counted electronically. The method was used for demonstration of the attractant activity of PDGF (see Table 1) and complex peptide mixtures such as proteose peptone and yeast extract. The assay was also used for studies on the chemoattractant effect of ß-endorphin (O'Neill et al. 1988) and other hormones (Köhidai et al. 1994). Pauli and Berger (1992), who studied the repellent activity of anilines and chlorophenols, used the method to study chemorepulsion by having the test substances in the inner chamber (Pauli and Berger 1992). Köhidai and coworkers used the method to determine the repellent effect of volatile oils (Köhidai et al. 1995). The capillary assay in this version is sensitive but rather time-consuming. When using this method, one should be aware of air bubbles at unwanted positions and/or outlet of liquid from the inner to the outer compartment.

The *two-phase assay* was first described in 1986 (Hellung-Larsen et al. 1986). As shown in Fig. 1, cells are placed on top of a bottom liquid phase made denser by addition of metrizamide (4–5% w/v). Although metrizamide is a weak repellent, it is a good choice. Gelatin, Ficoll, or sucrose do not work as well as metrizamide as density substance for various reasons which are discussed in the original paper. It was found that the metazoan signal peptides EGF, f-met-leu-phe and PDGF (used in the nano- to micromolar concentration range, see Table 1), as well as a number of other proteins, were attractants. Some of the amino acids like cysteine, methionine, and phenylalanine were active at the millimolar level. At that time, the method was not as reproducible as desired due to (1) technical difficulties in applying the cell suspension on top of the metrizamide phase and (2) due to day-to-day biological variations in the behavior of the unstimulated controls. Recently, however, we made a systematic study of the different physiological states of *Tetrahymena* (Hellung-Larsen et al. 1993). On the basis of these systematic studies in our own laboratory and also based on results on aero- and geotaxis obtained by us during a visit in the laboratory of Professor Donat-P. Häder, Erlangen, Germany, we succeeded better in defining the physiological parameters affecting the chemosensory response of *Tetrahymena* (Koppelhus et al. 1994a). We have now improved the reproducibility of the two-phase bioassay considerably. In particular, the application of an automated thermostated spectrophotometer for registration of the cell concentration of the lower metrizamide phase (Table 1) has increased the reproducibility, as well as made the method less laborious. The following parameters are of importance for the reproducibility of the chemosensory response (1) the growth of the cells before starvation, (2) the starvation medium (pH, salts, ionic strength), (3) the time of starvation, (4) the temperature during starvation, (5) the cell concentration, and (6) the oxygen supply during starvation, (7) the mechanical disturbance during washing of cells, (8) the cell concentration, and (9) the temperature during the assay. By using the two-phase assay we extended the list of attractants (also shown in Table 1) to other signal peptides namely insulin, FGF, and ß-endorphin whereas oxytocin and vasopressin had no effect (see Table 1). Recently, Köhidai and Csaba (1996) have found that oxytocin and vasopressin induced chemorepulsion of *Tetrahymena pyriformis*.

Using chemoattractant-signal-to-noise ratios from the two-phase assay (carried out on starved cells) as a measure for the chemoattractant potential, a hierarchy of attractant effects (potentials) can be listed as follows: proteose peptone $\gg$ amino acid mixture + PDGF > amino acid mix > single peptide hormones > single amino acids > control > repellents (Koppelhus et al.

1994b). Recently, we found that cells maintained in chemically defined medium containing amino acids (Szablewski et al. 1991) also respond to insulin when using the two-phase assay without prior washing or starvation. The chemosensory response in this case must be due to the capability of the cells to discriminate between different substances even when present at the same time, and also the ability to discriminate between different concentrations of the same signal peptide. This point will be further discussed below when describing adaptation as a cellular mechanism used by the cells during the chemoattraction.

The *filter paper assay* is a quantitative population assay of the Boyden-type, as it is based on the same principle as used to measure leukocyte chemotaxis (Boyden 1962). He measured leukocyte migration through a membrane filter by an apparatus with two chambers separated by a Millipore filter where cells respond to the attractant gradient established by diffusion through the filter. Although *Tetrahymena*, as a ciliate, normally swims freely in a liquid, it also has the potential to enter a paper membrane filter and squeeze itself through a membrane with the appropriate pore size. Using *Tetrahymena thermophila*, Whatman 3MM Chr paper is suitable in this respect and a number of other filter paper types are also useful but suboptimal. In the filter paper assay two compartments are separated by the filter paper membrane. The cell suspension is placed in one subcompartment and the attractant in the other (Leick et al. 1990). Using proteose peptone as a attractant and an optimal-responding cell suspension a signal-to-noise ratio of about 10 can easily be obtained at 28 °C using *T. thermophila*. The method is, however, less sensitive to "weak" attractants than liquid assays (e.g., the two-phase assay) illustrated by the fact that a mixture of the 20 amino acids gives a clear response in the latter but not in the filter paper assay. As the pore size of Whatman 3MM is around 10–20 μm versus the dimensions of the cell (20 × 40 μm), it is likely that the cells move through the filter by a high intensity response only induced by "strong" chemoattractants like proteose peptone, PDGF, and FGF (Leick et al. 1990, 1994).

The *field-assay* has primarily been used for more detailed analysis of single cellular behavior under specified conditions (Table 1, Fig. 2). The method is based on the principle in the leukocyte chemotactic chamber of Zigmond (1977). Various modifications have been used, but a general feature is that two chambers – one with cells and the other with attractant – are separated by a bridge (1–2 mm and with a height of 50–100 μm). Dramatic reduction of swimming speed can be achieved by placing semisolid media (gelatin or methylcellulose) on the bridge (Leick 1988; Leick et al. 1994). This also makes ciliary activity visually more conspicuous and makes it possible to monitor ciliary beat frequencies and to score the reversals of the methacronal ciliary waves (see below). The chamber is placed under a microscope with or without video recording, and single cellular behavior (swimming speed, angle, turnings, tumblings, idlings, etc.) can be monitored and subjected to computer-aided motion analysis. Using this assay we have analyzed cell accumulation and swimming speed of *T. pyriformis* as a function of distance from attractant. We confirmed (Hellung-Larsen et al. 1986) the results obtained by Levandowsky et al. (1984) that the swimming speed increases when the cells approach an attractant. Furthermore, we found that, at a distance 2–5 mm from the attractant, the cells moving towards the attractant have a higher speed than those swimming away from it (Leick et al. 1994).

The *capillary assay* (version 2), using single glass capillaries, is a simple and sensitive method, but it is semiquantitative (Almagor et al. 1981; Levandowsky et al. 1984). It was used by Koppelhus et al. (1994b) to judge the importance of the physiological parameters mentioned above. It was found that the most reproducible and sensitive chemosensory response was obtained by using cells starved at 21 °C for about 20–40 h. Furthermore, the size of the air/liquid surface of the starvation medium should be kept high to ensure proper oxygenation of the cells. Essentially identical results were obtained whether or not the cells were starved from late exponential growth phase or from the stationary phase. The optimal cell concentration during starvation was 2×10^5 cells/ml. The *T-maze assay* for chemokinesis of *Paramecium* brings a cell suspension in control solution loaded into the entry (E) arm of the T maze in contact with a test solution loaded into a T arm (Van Houten et al. 1975). Control solution fills the C arm and the plug. The stopcock is opened and cells swim into the plug, where they are presented with the test and control solution. After 30 min, cells in the E, C, and T arms are counted and the number of cells in T divided by the number of cells swimming into either T or C gives an index of chemokinesis. A similar kind of relative scoring of the results was used in the *capillary tube chemoresponse assay* (Francis and Hennessey 1995). In this assay, carried out in single glass capillaries, chemorepellent effects were assayed using *Paramecium tetraurelia* and *Tetrahymena thermophila*. Cytochrome c was found to be a repellent to *Paramecium* but not to *Tetrahymena*.

In the method using the so-called *double-P cutter*, channels are cut in agar layers and a diffusion gradient of attractant is established in the cuts. After incubation and fixation of the cells, the number of positively responding cells is determined in the container filled with attractant by stereomicroscopy in dark field. Complex Tryptone/yeast extract medium was an attractant by this method (Köhidai 1995).

3.2 Single Cell Assays for Chemoattraction

Since population assays measure the average behavior of cells in a cell suspension, these methods must be supplemented with studies of single cell behavior in order to obtain more detailed information about cellular behavioral mechanisms of single cells when stimulated with a chemoattractant. The literature contains a number of sophisticated computer-aided procedures to monitor ciliate behavior, see for example Häder and Lebert (1985).

The *field-assay* (modified Zigmond chamber) was used by us for studies on single cell locomotion of *Tetrahymena thermophila* in response to different peptides. Using 5–10% gelatin or methylcellulose on the bridge in the Zigmond chamber, evidence was provided that cells applied a spatial element in their locomotion (oriented chemokinesis) when migrating in semisolid media along a gradient of proteose peptone or PDGF (Leick 1988). The starved cells transform into elongated "gliding" cells which move about 2 orders of magnitude more slowly than under free-swimming conditions. Oriented locomotion of single cells was also shown under nonsemisolid condition, i.e., in gradients of proteose peptone (Hellung-Larsen et al. 1990). By computer-aided motion analysis, it was found that 65% of the cells are oriented within the 180° angle segment towards the attractant compared to 51% in the control, containing no attractant. Recently, a more detailed analysis was carried out on cells migrating in semisolid methylcellulose. It was found that the response of "gliding" cells, in gradients of attractants, was discontinuous. Cells stimulated by an attractant make time runs ("glide") between stops of locomotion lasting approx. 2 min. Moreover, stimulated cells migrating towards the attractant suppressed ciliary reversals less frequently than cells migrating in the opposite direction, thus making longer runs between stops. This adaptative behavior of the cells will be discussed further below (Leick et al. 1994).

3.3 Assays of Ciliary Activity

Normal ciliary activity during swimming (ciliary metachronal waves) of "gliding" cells is microscopically conspicuous at rather low magnification. It is possible visually to follow, fairly easily, ciliary waves commencing at the very anterior cellular tip and running approx. one third down the long axis of the cell. Ciliary waves in the posterior part of the cell can also be seen at high magnifications, although they are less conspicuous. The frequency of the anterior ciliary waves can be measured on sequences of selected still frames of video film recordings and/or on the frame-grabbed stills from digital video recordings made by a computer. Analysis of cells "gliding" in 5–10% methylcellulose (v/w) with ciliary waves beating towards the posterior end of the cell showed these frequencies (by a semiquantitative estimate) to be in the range 2–5 Hz (Leick et al. 1994). The high viscosity of the methylcellulose probably contributes to decreasing the metachronal beat frequency compared to that of free-swimming cells. A combination of a high concentration of cilia at the cellular anterior and the high viscosity of the surrounding medium is probably the major reason for cell elongation during "gliding". Ciliary reversal of the active ciliary stroke results in an immediate withdrawal of the cellular tip and makes it possible to determine time runs of individual cells. A time run was operationally defined as terminated when any ciliary reversal lasted for more than 10 s. Using this method, it was possible to estimate that unstimulated control cells spend about the same amount of time in the phase with normal ciliary beating (a time run) as in the ciliary reversed stage (a stop), i.e., around 2 min. In contrast, proteose peptone stimulated cells have time runs of 11.8 ± 3.9 min before reversing their active ciliary stroke for more than 10 s. The stimulus-dependent prolongation of the swimming stage I ("gliding") was followed by a return (adaptation) to the level of the ciliary reversal frequency seen before the attractant was added (control level) under conditions when the concentration was kept constant (no gradient).

4 Cellular Mechanisms Related to Peptide Action on Individual Cell Behavior

4.1 Adaptation

Adaptation is a widespread feature of many chemotactic cells and other sensory systems and is a kind of short-term memory (Morimoto and Koshland 1991). Adaptation is here defined as a reversible elimination of the responsiveness of a cell caused by an adjustment in the sensitivity of the cell (Zigmond and Sullivan 1979). An adapted organism responds to a change in stimulus intensity and ceases to respond in the absence of any further change. Adaptation permits a sesory system to operate with constant efficiency over a large range of stimulus intensities, that is, during migration along a gradient of an attractant. In fact, a computer simulation study has confirmed that the altered frequency with which the ciliate *Paramecium* changes direction in combination with adaptation can bring about cellular aggregation, but in the absence of adaptation it is ineffective (Van Houten and Van Houten 1982).

The involvement of adaptive behavior in chemoattraction was suggested when it was observed that cells kept in defined growth medium were attracted to proteose peptone or PDGF (Leick and Hellung-Larsen 1985). A more systematic study of the swimming behavior of individual cells in gradients of the chemoattractant proteose peptone (or FGF) revealed that cells adapt even to ambient individual concentrations of each substance. Since the cells make fewer turns (smoother swimming) when swimming towards higher concentrations of an attractant than when swimming towards lower concentrations, the net movement is positive towards the attractant. The altered frequency of turns in response to the change in concentration is a transient response; after a time, the cells adapt to the new concentration and the turning frequency returns to a normal, or baseline, level (Leick et al. 1994).

The concentration range of proteose peptone in which the cells are able to respond varies from approx. 1 µg/ml up to 20 mg/ml. It should be noted that cells are able to adapt to a new ambient level of proteose peptone even when this attractant (0.2 mg/ml) is placed directly in the liquid cell suspension in the Zigmond chamber. If they did not adapt/deadapt to 0.2 mg/ml proteose peptone, they would be unable to respond to the gradient in the gel. Thus, adaptation to an attractant is not dependent on the viscosity of the medium or on whether or not the cells, which enter the attractant gradient, are starved or not starved (Leick et al. 1994). Adaptation is most clearly revealed when analyzing cell behavior of individual cells in the modified Zigmond chamber, a small fraction of "gliding" cells are seen to move away from the attractant (Fig. 2). These cells swim less smoothly, meaning that they are stimulated with respect to their ciliary reversals, when compared to control cells, and they behave like cells reacting to a repellent with increased frequency of ciliary reversals. Using free-swimming cells and proteose peptone as attractant, cells have only a few seconds available to adapt to a particular ambient concentration of attractant when swimming on the bridge in the Zigmond chamber. In fact, when cells are attracted to the gel before entering the semisolid phase of the Zigmond chamber (Fig. 2), adaptation to the attractant concentration at the edge of the gel is probably the most important element leading to accumulation of free swimmers at this edge. Free-swimming adapted cells trying to escape the attractant at the edge are repelled (less smooth swimming) when they try to swim back into the free solution. This leads to

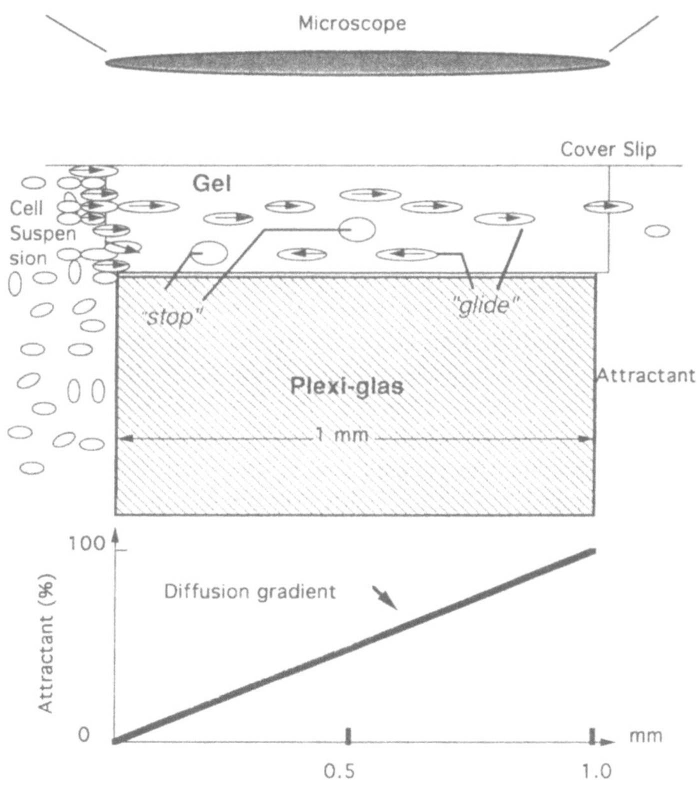

Fig. 2. Schematic drawing of single cell behavior in semisolid 5–10% methylcellulose in the field assay in modified Zigmond chamber

accumulation of a cloud of cells at the gelified attractant (Fig. 2). Preliminary video recordings of cell tracks of free swimmers reacting to a gradient of proteose peptone without solidifying substance in the Zigmond chamber confirm that cells swimming towards lower attractant concentrations are repelled (Leick, unpubl. experiments). Accordingly, it is presently not necessary, regarding adaptation, to formulate a different type of cellular mechanisms for the adaptation of free-swimming cells and slow "gliders" migrating in semisolid media in the Zigmond chamber (Fig. 2).

In *Paramecium* the chemokinetic response certainly involves adaptation as discussed in the theoretical paper by Lapidus and Levandowsky (1981). The term habituation which is used in studies of animal behavior is synonymous to adaptation as used here.

4.2 Persistence

When using different bioassays it was consistently observed that peptide-attractants work with different intensities, depending on their chemical nature. When, for example, one is using chemoattractant signal-to-noise ratios from the two-phase assay as a measure for the chemoattractant potential, a hierarchy of attractant effects (potentials) was listed as discussed above. When using the field assay with a gel or the filter paper assay, where there is physical hindrance for the cells to overcome in order to be scored as responders, only "strong" attractants will work. When cells are migrating through semisolid gelatine in the Zigmond chamber, only proteose peptone, FGF, and PDGF are strong enough chemoattractants to induce oriented cell "gliding" under these viscous conditions (Leick et al. 1994). In the paper membrane assay (Boyden-type assay) developed for *Tetrahymena* described above, only proteose peptone and PDGF were able to induce altered locomotion strong enough to make the cells swim through a filter paper membrane (Leick et al. 1990). The high-intensity responses have recently been further characterized as an attractant-induced persistence of cellular swimming caused by intensive ciliary beating leading to much longer time runs for cells swimming positively towards an attractant than for cells swimming negatively away from an attractant (Leick et al. Addendum, *Note 1*). It was also shown that the plant lectin concanavalin A (ConA) specifically interferes with chemoattraction during high-intensity responses, that is, persistent ciliary beating in semisolid substances. Cell surface receptors for ConA (FITC-ConA) were seen to cluster at the front of the cells, suggesting that a high concentration of mannose-containing receptors at the cellular anterior is important for persistent cell migration. This persistence in time of the active ciliary strokes (metachronal waves) is a major cellular mechanism for a directional "memory" during chemoattraction in semisolid substances. In contrast, no effect of ConA was found on low-intensity responses involving strong chemoattractants in all the liquid assays where the physical hindrance that the cells have to overcome is minimal. In liquids, i.e., capillary assays or the two-phase assay, persistent swimming is less important, and adaptation is probably here the major cellular mechanism responsible for chemoattraction.

5 Concluding Remarks

The general pattern for chemoattraction in protists, including the free ciliates, seems to be that signal molecules help to define ecological niches important for a particular biological function for the survival of a particular organism, thereby making it possible for different microorganisms to coexist in the same environment but still maintain their identity. In the case of *Tetrahymena*, peptides are important signal molecules for accumulating cells by chemoattraction in a microenvironment which signals food and proliferation and survival in that particular microenvironment. The involvement of signal substances in survival and the promotion of the growth process has been reported for insulin (Christensen 1993; Grave et al. 1995; Kristiansen et al., in prep.).

It is puzzling why a series of mammalian endocrine peptide hormones are efficient chemoattractants for *Tetrahymena*. They exert their effect in low hormonal concentrations as in the endocrine system (Table 1). This leads to the question of whether or not the particular receptors are present in the plasma membrane of *Tetrahymena*.

Csaba, who for many years has worked on *Tetrahymena* as a model system for cell receptor research and on development of hormone receptors, has suggested that cell-membrane-associated signal receivers in this organism may be transient patterns arising by continuous dynamic change of the cell membrane (Csaba 1985). He has proposed that the fluid mosaic membrane already present at the unicellular level makes possible not only movement of membrane proteins, but also their assembly in different configurations, making it possible dynamically to change the receptor configuration (Koch et al. 1979).

The fact that the opiate receptor which binds the chemoattractant ß-endorphin has been isolated and partially characterized in *Tetrahymena pyriformis* as a 110-kDa protein similar to the receptor found in rat brain and the leech (O'Neill et al. 1988) indicates that classical receptor proteins for peptide hormones certainly are present in the *Tetrahymena* plasma membrane (Zipser et al. 1988; Addendum, *Note 2*). *Tetrahymena thermophila* is sensitive to ß-endorphin (and α-endorphin) and to met-enkephalin as well (Koppelhus et al. 1994b), indicating that these hormones induce chemoattraction in *Tetrahymena* as in blood mononuclear cells (Van Epps and Saland 1984).

In most of the experiments where chemosensory and other biological effects of peptides and proteins have been tested on *Tetrahymena*, either *T. thermophila* or *T. pyriformis* was used. According to the rRNA sequence analysis of these two species, they are rather distantly related (Preparata et al. 1989). Although there are differences, they behave rather similarly in their chemosensory responses. However, it would be of interest to make comparative studies between different species.

Evidence for the presence of an opioid receptor has also been provided from studies on *Amoeba proteus* (Josefsson and Johansson 1979). LeRoith and Roth proposed that peptide hormone signal recognition (and transduction) developed rather early in evolution (LeRoith and Roth 1984). Nanney has suggested that ciliates were probably the first wave of organisms that exploited the eukaryotic machinery in lakes and seas some 2 billion years ago (Nanney 1982). However, they may also have been the first organisms to develop a peptide-signal/receptor system for the purpose of accumulating (and grow) in a suitable microenvironment, which may have included a behavioral system by which they were attracted to each other. In very dense old shallow cultures of *Tetrahymena*, regular patterns of swimming cells sometimes form that appear perhaps to be due to chemosensory behavior (Levandowsky and Hauser 1978).

In David Nanney's paper *Genes and Phenes*, he emphasizes the evolutionary conservatism of the phenotype of ciliates in contrast to the genotype which had the opportunity to change within the constraints dictated by the conservation of its fundamental phenotype (Nanney 1982). It is therefore possible that the development of peptide-effector/receptor systems has shown the same conservatism during evolution as ribosomes, nucleosomes, and cilia, and that *Tetrahymena* also in this respect must be considered a true "living fossil", meaning that the striking similarities of the chemoattractant system of this ciliate and the peptide hormonal signaling system in vertebrates allows a suggestion that these lower eukaryotes may well have been the evolutionary precursors of the latter.

Acknowledgment. The excellent technical assistance of Mrs. Charlotte Iversen and Mrs. Inger Lyhne is gratefully acknowledged.

Addendum. The following notes added in proof are information obtained after submission of the manuscript.

Note 1. Symposium contribution by V. Leick in: Preston TM, Grebecki A. Protistological Actualities. Proceedings of the Second Congress of Protozoology Clermont-Ferrand 1995. Eds: G. Brugerrolle and J.-P. Mignot. Symp. 10: 185–194

Note 2. Christopher and Sunderman reported recently about the isolation of an insulin-like ciliary membrane protein in *Tetrahymena pyriformis* (62–67 kDa) that may function as both a precursor and a receptor for insulin. Christopher GK and Sundermann CA (1995) Biochem Biophys Res Commun 212: 515–523.

References

Almagor M, Ron A, Bar-Tana J (1981) Chemotaxis in *Tetrahymena thermophila.* Cell Motility 1: 261–268

Andersen HA, Flodgaard H, Klenow H, Leick V (1984) Platelet-derived growth factor stimulates chemotaxis and nucleic acid synthesis in the protozoan *Tetrahymena.* Biochim Biophys Acta 782: 437–440

Bailey GB, Leitch GJ, Day DB (1985) Chemotaxis by *Entamoeba histolytica.* J Protozool 32: 341–346

Blum JJ (1967) An adrenergic control system in *Tetrahymena.* Proc Natl Acad Sci USA 58: 81–88

Boyden S (1962) The chemotactic effect of mixtures of antibody and antigen on polymorphonuclear leucocytes. J Exp Med 115: 453–466

Christensen ST (1993) Insulin rescues the unicellular eukaryote *Tetrahymena* from dying in a complete, synthetic nutrient medium. Cell Biol Int 17: 833–837

Christensen ST, Rasmussen L (1992) Evidence for growth factors which control cell multiplication in *Tetrahymena thermophila*. Acta Protozool 31: 215–219

Christopher GK, Sundermann CA (1992) Conventional and confocal microscopic studies of insulin receptor induction in *Tetrahymena pyriformis*. Exp Cell Res 201: 477–484

Corliss JO (1979) The ciliated protozoa, 2nd edn. Pergamon Press, Oxford

Csaba G (1985) The unicellular *Tetrahymena* as a model cell for receptor research. Int Rev Cytol 95: 327–377

Csaba G, Kovács P (1992) Oxytocin and vasopressin change the activity of the contractile vacuole in *Tetrahymena*: newer contributions to the phylogeny of hormones and hormone receptors. Comp Biochem Physiol 102A: 353–355

Csaba G, Lantos T (1975) Effect of insulin on the glucose uptake of protozoa. Experientia 31: 1097–1098

De Jesus S, Renaud FL (1989) Phagocytosis in *Tetrahymena thermophila*: naloxone-reversible inhibition by opiates. Comp Biochem Physiol 92C: 139–142

De Wit RJW, van Bemmelen MXP, Penning LC, Pinas JE, Calandra TD, Bonner JT (1988) Studies of cell-surface glorin receptors, glorin degradation, and glorin-induced cellular responses during development of *Polysphondylium violaceum*. Exp Cell Res 179: 332–343

Francis JT, Hennessey TM (1995) Chemorepellents in *Paramecium* and *Tetrahymena*. J Euk Microbiol 42: 78–83

Gerisch G (1987) Cyclic AMP and other signals controlling cell development and differentiation in *Dictyostelium*. Annu Rev Biochem 56: 829–852

Grave M, Hagemeister JJ, Kristiansen TB, Lyhne I, Hellung-Larsen P (1995) Growth factor removal from *Tetrahymena* cells leads to cell death at the medium-air interphase. Proc 4th Asian Conf Ciliate Biol, Tokyo

Häder D-P, Lebert M (1985) Real time computer-controlled tracking of motile microorganisms. Photochem Photobiol 42: 509–514

Hellung-Larsen P, Leick V, Tommerup N (1986) Chemoattraction in *Tetrahymena*: on the role of chemokinesis. Biol Bull 170: 357–367

Hellung-Larsen P, Leick V, Tommerup N, Kronborg D (1990) Chemotaxis in *Tetrahymena*. Eur J Protistol 25: 229–233

Hellung-Larsen P, Lyhne I, Andersen AP, Koppelhus U (1993) Characteristics of dividing and non-dividing *Tetrahymena* cells at different physiological states. Europ J Protistol 29: 182–190

Holz GG Jr (1973) The nutrition of *Tetrahymena*: essential nutrients, feeding and digestion. In: Elliott AM (ed) Biology of *Tetrahymena*. Dowden, Hutchinson & Ross, Stroudsburg, Pennsylvania, pp 89–98

Honda H, Miyake A (1975) Taxis to a conjugation-inducing substance in the ciliate *Blepharisma*. Nature 257: 678–680

Janakidevi K, Dewey VC, Kidder GW (1966) The biosynthesis of catecholamines in two genera of protozoa. J Biol Chem 241: 2576–2578

Josefsson J-O, Johansson P (1979) Naloxone-reversible effect of opioids on pinocytosis in *Amoeba proteus*. Nature 282: 78–80

Koch AS, Fehér J, Lukovics I (1979) Single model of dynamic receptor pattern generation. Biol Cybern 32: 125–138

Köhidai L (1995) Method for determination of chemoattraction in *Tetrahymena pyriformis*. Curr Microbiol 30: 251–253

Köhidai L, Csaba G (1996) Different and selective chenmotactic responses of *Tetrahymena pyriformis* to two families of signal molecules: lectins and peptide hormones. Acta Microbiol Immunol Hung 43: 83–91

Köhidai L, Barsony J, Roth J, Max SJ (1992) Rapid effects of insulin on cyclic GMP location in an intact protozoan. Experientia 48: 476–481

Köhidai L, Karsa J, Csaba G (1994) Effects of hormones on chemotaxis in *Tetrahymena*: investigations on receptor memory. Microbios 77: 75–85

Köhidai L, Lemberkovics E, Csaba G (1995) Molecule dependent chemotactic responses of *Tetrahymena pyriformis* elicited by volatile oils. Acta Protozool 34: 181–185

Koppelhus U, Hellung-Larsen P, Leick V (1994a) Physiological parameters affecting the chemosensory response of *Tetrahymena*. Biol Bull 187: 1–7

Koppelhus U, Hellung-Larsen P, Leick V (1994b) An improved quantitative assay for chemokinesis in *Tetrahymena*. Biol Bull 187: 8–15

Kovács P, Csaba G, Nagao S, Nozawa Y (1989) The regulatory role of calmodulin-dependent guanylate cyclase in association with hormonal imprinting in *Tetrahymena*. Microbios 59: 123–128

Lapidus IR, Levandowsky M (1981) Mathematical models of behavioral responses to sensory stimuli by protozoa. In: Levandowsky M, Hutner SH (eds) Biochemistry and physiology of protozoa, 2nd edn, vol 4. Academic Press, New York, pp 235–260

Leick V (1988) Gliding *Tetrahymena thermophila*: oriented chemokinesis in a ciliate. Eur J Protistol 23: 354–360

Leick V (1992) Chemotactic properties, cellular binding and uptake of peptides and peptide derivatives: studies with *Tetrahymena thermophila*. J Cell Sci 103: 565–570

Leick V, Helle J (1983) A quantitative assay for ciliate chemotaxis. Anal Biochem 135: 466–469

Leick V, Hellung-Larsen P (1985) Chemosensory responses in *Tetrahymena*: the involvement of peptides and other signal substances. J Protozool 32: 550–553

Leick V, Hellung-Larsen P (1992) Chemosensory behaviour of *Tetrahymena*. BioEssays 14: 61–66

Leick V, Frederiksen K, Lyhne I, Hellung-Larsen P (1990) A paper membrane filter assay for ciliate chemoattraction. Anal Biochem 184: 63–66

Leick V, Koppelhus U, Rosenberg J (1994) Cilia-mediated oriented chemokinesis in *Tetrahymena thermophila*. J Euk Microbiol 41: 546–553

LeRoith D, Roth J (1984) Vertebrate hormones and neuropeptides in microbes: evolutionary origin of intercellular communication. In: Martini L, Ganong WF (eds) Frontiers in neuroendocrinology, vol 8. Raven Press, New York, pp 1–25

LeRoith D, Delahunty G, Wilson GL, Roberts CT Jr, Shemer J, Hart C, Lesniak MA, Shiloach J, Roth J (1986) Evolutionary aspects of the endocrine and nervous systems. Recent Prog Horm Res 42: 549–587

Levandowsky M, Hauser DCR (1978) Chemosensory responses of swimming algae and protozoa. Int Rev Cytol 53: 145–210

Levandowsky M, Cheng T, Kehr A, Kim J, Gardner L, Silvern L, Tsang L, Lai G, Chung C, Prakash E (1984) Chemosensory responses to amino acids and certain amines by the ciliate *Tetrahymena*: a flat capillary assay. Biol Bull 167: 322–330

Luporini P, Miceli C (1986) Mating pheromones. In: Gall JG (ed) The molecular biology of ciliated protozoa. Academic Press, New York, pp 263–299

Lwoff A (1923) Sur la nutrition des infusoires. C R Acad Sci 176: 928–930

Miyake A (1981) Cell interaction by gamones in *Blepharisma*. In: O'Day DH, Horgen PA (eds) Sexual interactions eukaryotic microbes. Academic Press, New York, pp 95–129

Morimoto BH, Koshland DE Jr (1991) Short-term and long-term memory in single cells. FASEB J 5: 2061–2067

Nanney DL (1982) Genes and phenes in *Tetrahymena*. BioScience 32: 783–788

Ortenzi C, Miceli C, Bradshaw RA, Luporini P (1990) Identification and initial characterization of an autocrine pheromone receptor in the protozoan ciliate *Euplotes raikovi*. J Cell Biol 111: 607–614

O'Neill JB, Pert CB, Ruff MR, Smith CC, Higgins WJ, Zipser B (1988) Identification and characterization of the opiate receptor in the ciliated protozoan, *Tetrahymena*. Brain Res 450: 303–315

Pauli W, Berger S (1992) Chemosensory and electrophysiological responses in toxicity assessment: investigations with a ciliated protozoan. Bull Environ Contam Toxicol 49: 892–899

Preparata RM, Meyer EB, Preparata FP, Simon EM, Vossbrinck CR, Nanney DL (1989) Ciliate evolution: the ribosomal phylogenies of the tetrahymenine ciliates. J Mol Evol 28: 427–441

Preston RR, Usherwood PNR (1988) Characterization of specific L-[3]-glutamic acid binding site on cilia isolated from *Paramecium tetraurelia*. J Comp Physiol B 158: 345–351

Renaud FL, Chiesa R, De Jesús JM, López A, Miranda J, Tomassini N (1991) Hormones and signal transduction in protozoa. Comp Biochem Physiol 100A: 41–45

Roth J, LeRoith D, Lesniak MA, de Pablo F, Bassas L, Collier E (1986) Molecules of intercellular communication in vertebrates, invertebrates and microbes: do they share common origins? Progr Brain Res 68: 71–79

Soldo AT, Van Wagtendonk WJ (1969) The nutrition of *Paramecium aurelia*, stock 299. J Protozool 16: 500–506

Szablewski L, Andreasen PH, Tiedtke A, Florin-Christensen J, Florin-Christensen M, Rasmussen L (1991) *Tetrahymena thermophila*: growth in synthetic nutrient medium in the presence and absence of glucose. J Protozool 38: 62–65

Van Epps DE, Saland L (1984) *β*-endorphin and met-enkephalin stimulate human peripheral blood mononuclear cell chemotaxis. J Immunol 132: 3046–3053

Van Houten J (1978) Two mechanisms of chemotaxis in *Paramecium*. J Comp Physiol A 127: 167–174

Van Houten J (1992) Chemosensory transduction in eukaryotic microorganisms. Annu Rev Physiol 54: 639–663

Van Houten J (1994) Chemosensory transduction in eukaryotic microorganisms: trends for neuroscience? Trends Neurosci 17: 62–71

Van Houten J, Van Houten JC (1982) Computer analysis of *Paramecium* chemokinesis behavior. J Theor Biol 98: 453–468

Van Houten J, Hansma H, Kung C (1975) Two quantitative assays for chemotaxis in *Paramecium*. J Comp Physiol 104: 211–223

Zigmond SH (1977) Ability of polymorphonuclear leucocytes to orient in gradients of chemotactic factors. J Cell Biol 75: 606–616

Zigmond S, Sullivan SJ (1979) Sensory adaptation of leucocytes to chemotactic peptides. J Cell Biol 82: 517–527

Zipser B, O'Neill JB, Ruff MR, Smith CC, Higgins WJ, Pert CB (1988) The opiate receptor: a single 110 kDa recognition molecule appears to be conserved in *Tetrahymena*, leech and rat. Brain Res 463: 296–304

Ciliate Pheromones

P. Luporini, C. Miceli, C. Ortenzi, and A. Vallesi[1]

1 Introduction

Our understanding of the biology of ciliate pheromones, and of the primary functions of the mating-type mechanism regulated by these molecules, has given rise to radical conflicts of opinion among different investigators, as regards interpretation of phenomena, definition of basic concepts, and use of terminology. Such conflicts are largely due to untimely generalizations of principles and to definitions not adequately based on a sufficiently wide spectrum of experimental observations. Over the years, we have relied mainly on studies on *Paramecium* and *Blepharisma*, which cannot be regarded in this context as paradigmatic systems of ciliates, and these latter form a rather heterogenous group of organisms (Schlegel 1991). It is, therefore, necessary to summarize concisely the historical background against which our understanding of ciliate pheromones has developed, in such a way as to provide the reader with a picture of the facts which is as coherent and objective as possible.

2 Background

A number of ciliate species exploit sex in the formation of pairs of mating, or conjugating, cells (for a recent review, see Dini and Nyberg 1993). Some of these species have evolved genetic systems controlling the manifestation of this mating process; others, such as some species of heterotrichs, commonly referred to as selfers, lack such a system and undergo mating unpredictably, without any apparent rules. As first found by Sonneborn (1937) in *Paramecium aurelia*, now denoted *P. primaurelia*, such a control of mating implies that the cells have the ability to discriminate between sibs and nonsibs (or self and nonself), and hence that at least two chemically distinct cell phenotypes may be present in one biological species. The discovery of these phenotypes was based on the observation that to elicit pair formation, mixing cell lines of different

[1]Dipartimento di Biologia Molecolare, Cellulare e Animale, Università di Camerino, 62032 Camerino (MC), Italy

lineages was as a rule necessary; mixing cells of the same lineage did not usually result in mating. Conspecific cell lines which, on mixing, undergo mating were said to belong to distinct mating types, and the putative determinant molecules were called mating-type substances.

Ciliate mating types, in spite of their designation, are, however, *not* equivalent to the complementary cell types denoted in other systems as sexes (Beale 1954; Nanney 1980; Luporini and Miceli 1986; Dini and Nyberg 1993). Mating types are properties of vegetative (asexual) cells, whose life is regulated by a polyploid somatic *macro*nucleus and which contain also a transcriptionally inert, diploid, germinal *micro*nucleus. The rejection of the analogy of ciliate mating types with sex is supported by the following facts. First, since the pioneering studies on *Euplotes patella* by Kimball (1942), it has been known that there may be more than two mating types within a single species; in the hypotrichs, the number of such mating types seems to be unlimited, as shown by the fact that when new strains are collected and monitored by their mating reactions with reference strains, new mating types are usually found (see, for example, Caprette and Gates 1994). Secondly, mating pairs may be *heterotypic*, i.e., consist of cells of two *different* mating types, or *homotypic*, i.e., involve cells of the *same* mating type; and the developmental processes which heterotypic and homotypic pairs (and their descendants) undergo are alike. A third fact opposing the supposed equation of mating types with sexes is the presence of both male (migratory) and female (stationary) gamete nuclei within every mating ciliate cell, irrespective of mating type. More generally, it may be said that if a ciliate cell is in a physiological state favorable for mating, it may mate (and subsequently exchange cytoplasm and genes) with *any* other conspecific cell that is also in a state favorable for mating, and that this process is independent of the specificity of the stimulus initiating the mating process.

These considerations have in the past been largely neglected, due to the fact that most mating-type studies have been carried out with *Paramecium* species, which possess basically *bipolar* mating-type systems. Furthermore, the mating-type substances of *Paramecium* are not secreted into the surrounding medium, but remain bound to the surface membranes of the cells (see, as reviews, Hiwatashi 1988; Kitamura 1988; Watanabe 1990). Consequently, mating-type systems of ciliates have been previously interpreted (functionally and terminologically) in terms of bipolar sexual systems. Two cells expressing different mating types have been said to be "complementary for fertilization" (Miyake 1978, 1981a) and the mating-type substances have been considered to be "sexual" factors which cells produce in order to develop a mutual "stimulation" causing readiness to mate.

Experimental support for these views has in the past been thought to be provided by the isolation and chemical identification of two mating-inducing molecules from culture filtrates of *Blepharisma japonicum*. One such substance was identified as a basic (pI = 7.5) glycoprotein of about 20 kDa (Miyake and Beyer 1974), the other as a 3-(2′-formylamino-5′-hydroxybenzoyl)lactate

(Kubota et al. 1973), which is a possible tryptophan derivative (Jaenicke 1984). These two substances were assigned to one or other of two mutually complementary mating types (I and II) of *Blepharisma* (Miyake and Beyer 1974) although, as earlier noted by Luporini and Miceli (1986), no genetic evidence bearing on the determination of these mating types has been obtained. These substances were called (G1 and G2) gamones, a term which means gamete hormones, by analogy with the mating substances described by Hartmann and Schartau (1939) in the green alga *Chlamydomonas*. It was then proposed that the G1 gamone of mating-type I cells exerted its mating effect by binding to specific hypothetical receptors carried *only* by cells of mating type II, and that G2 gamone exerted its mating effects by binding to specific receptors carried *only* by cells of mating-type I.

This gamone-receptor hypothesis of *Blepharisma* was later applied to other ciliates, including those exhibiting multiple mating-type systems (Miyake 1981b). It was then postulated that, within a group of cells belonging to a set of mutually compatible mating types, each gamone-secreter cell would carry all types of receptors specific for every gamone in the set, except those specific for the cell's "own" gamone(s). As an example, we may consider a hypothetical situation with four mating types (denoted I–IV), each type producing a corresponding gamone (G1-G4, respectively). Cells of mating type I producing G1 are assumed to carry receptors R2, R3, and R4, but not R1; similarly, cells of mating type II producing G2, would carry R1, R3, and R4, but not R2; and so on. Mating in a mixture between cells of mating types I and II would involve formation of the complexes R1 + G1 and R2 + G2 on the respective cell surfaces.

As pointed out by Beale (1990), difficulties in the inferences and predictions resulting from this gamone-receptor hypothesis became apparent when work was done on two new species of *Euplotes*, *E. octocarinatus* and *E. raikovi*, which have been studied in the laboratories of K. Heckmann (Münster) and P. Luporini (Camerino), respectively. These ciliates, like *E. patella* (Kimball 1942; Akada 1985), secrete soluble mating-inducing substances controlled by series of codominant alleles at single loci denoted *mat* or *mt* (Akada 1986; Luporini et al. 1986; Kuhlmann and Heckmann 1989). Such *Euplotes* species may thus be regarded as more suitable systems than *B. japonicum* for the study of the control and activity of mating-type-dependent chemical signaling in ciliates. Results obtained from studies of *E. octocarinatus* have been interpreted in terms of the gamone-receptor hypothesis (Heckmann and Kuhlmann 1986; Kuhlmann and Heckmann 1989), while evidence from studies with *E. raikovi* has been thought to conflict with this hypothesis, and to support an alternative one, described as the "self-recognition hypothesis" (Luporini and Miceli 1986). This assumes that ciliate mating-type substances have the basic function to enable cells to discriminate self from nonself, and that their primary target consists of specific autocrine membrane receptors which are borne by the same cells as those from which the mating-type substances are produced. In this new context, the term gamone is inappropriate because of its original concern with "substances for interaction between cells complementary for fertilization" (Miyake 1983). It was thus proposed to replace the word gamone with "mating pheromone" (Luporini and Miceli 1986), or simply pheromone, which is the term now in use.

Progress in the knowledge of the biology of these molecules is reviewed in this chapter, special attention being given to physiological, structural, and genetic aspects that have major implications for their function and mode of action.

3 Pheromone Notation and Origin

Pheromones of *E. raikovi* are usually denoted by the common abbreviation E*r*, where E stands for euplomone (i.e., *Euplotes* pheromone) and *r* for *raikovi*. Each pheromone is then identified by a number, which is the same as that used to identify the corresponding coding (*mat*) gene and mating type (hereafter referred to only as type) of the cell. This direct numerical correspondence between pheromones, genes, and cell types is also applied to cells heterozygous for two codominant *mat* genes producing two pheromones, in such a way that they are marked by the same two numbers as those used to indicate the corresponding homozygotes. In Table 1, this classification is shown with reference to the only *mat*-heterozygous wild strains from which we have obtained the *mat*-homozygous progeny clones that have been so far used to purify and characterize the *E. raikovi* pheromones. They are only some of a wider collection of strains (isolated from different localities of the Mediterranean and Western USA coasts) that represent other potential sources of new pheromones.

Breeding analyses of the *E. raikovi* strains (denoted 13, 19, 27, 1N, 3N, GA-2, and GA-4) so far used in the study of pheromones have shown them to

Table 1. *Euplotes raikovi* strains used for pheromone studies

Strain designation	Collecting sites	Secreted pheromones	Genotypes	Cell (mating) types
13	Porto Recanati[a]	E*r*-1; E*r*-2	*mat*-1/*mat*-2	I-II
19	"	E*r*-2; E*r*-7	*mat*-2/*mat*-7	II-VII
27	"	E*r*-3; ND[c]	*mat*-3/ND	III-ND
1N	"	E*r*-9; E*r*-10	*mat*-9/*mat*-10	IX-X
3N	"	E*r*-10; E*r*-11	*mat*-10/*mat*-11	X-XI
GA-2	Gaeta[b]	E*r*-21; ND	*mat*-21/ND	XXI-ND
GA-4	"	E*r*-20; ND	*mat*-20/ND	XX-ND

[a]Eastern, or Adriatic coast of Italy.
[b]Western, or Tyrrhenian coast of Italy.
[c]Not determined.

represent different cell types which constitute a substantial gene pool (Valbonesi et al., ms in prep.). However, the patterns of mating reactions, as they are usually detected in species with high-multiple mating-type systems, have revealed specific restrictions to a general mutual mating compatibility (Valbonesi et al. 1992; Caprette and Gates 1994). While mating occurs in every pairwise combination between strains 13, 19, 27, 1N, and 3N, and in the combination between strains GA-2 and GA-4, it does not occur in the other combinations between strains of the two groups. Therefore, pheromones E*r*-1, E*r*-2, E*r*-3, E*r*-7, E*r*-9, E*r*-10, and E*r*-11 which are secreted from one group of strains may be regarded as potential members of a subfamily distinct from another one including pheromones E*r*-20 and E*r*-21 secreted from a second group.

A different notation system has been adopted for *E. octocarinatus* pheromones; it follows the traditional system earlier used in *E. patella.* The four pheromones so far identified from breeding analyses of two wild strains isolated from the same freshwater pond in Germany (Heckmann and Kuhlmann 1986) are denoted G1-G4 and their coding gene as mt^1-mt^4, respectively. The ten cell types (four homozygotes and six heterozygotes) generated by the ten possible combinations of the four codominant genes are numbered from I to X, regardless of the pheromone and gene numbering.

4 Pheromone Secretion and Purification

Pheromones are continuously secreted by the cells and thus apparently follow a constitutive pathway of secretion. Immunocytochemical analyses have provided evidence that, in *E. octocarinatus*, on the way to the extracellular environment, they transit, and probably also accumulate in saccule-like structures (known as ampules) that are permanently associated with the cell plasma membrane lining the pits into which the ciliary organelles are inserted (Kusch and Heckmann 1988). Because of the traditional view that pheromones function only in inducing cell mating, it has commonly been held that pheromone secretion is a property only of cells which are in developmental and physiological conditions compatible with mating, i.e., cells that have reached the adult (or mature) stage of their life cycle, and are entering starvation after food depletion. However, quantitative analyses of secreted pheromones and pheromone gene transcripts carried out in *E. raikovi* have provided direct evidence that pheromone synthesis and secretion occur independently of the cell mating competence (Luporini et al. 1992). They in fact occur, as shown in Tables 2 and 3, also in cells which are at stages incompatible with mating, such as those at the beginning of the clonal life cycle (defined as immaturity), and also in well-fed cells.

Different procedures have been adopted for pheromone purification in *E. raikovi* and *E. octocarinatus*. In the former, pheromones are adsorbed from cell

Table 2. Production of E*r*-2 mRNA and secreted pheromone E*r*-2 in type II cells of *E. raikovi* at different stages of the life cycle

Cell age (postconjugation fissions)	E*r*-2 mRNA (%)	Secreted E*r*-2 (μg/l)
5–7 (Immaturity)	17	3.68
20–30 (Immaturity)	25	4.96
> 100 (Maturity)	100	51.6

For RNA analysis, cells of three different ages were harvested, washed, and resuspended in fresh seawater for 4 h before total RNA extraction. RNA samples were probed, in a dot-blot analysis, with the plasmid insert containing the coding region of preproE*r*-2 (Miceli et al. 1991; Luporini et al. 1992). The amount of hybridization was estimated as radioactivity, and E*r*-2 mRNA was quantified as percentages (%) of the maximum value of radioactivity measured at the maturity stage.
Secreted E*r*-2 was measured by anion exchange chromatography on a Mono Q HR 5/5 column using 1-l volumes of filtrates from cell cultures suspended at the concentration of 10^4/ml. Values are means of two independent measurements.

Table 3. Isolation of pheromone E*r*-1 from filtrates of fed cultures of *E. raikovi* cells

Time (h)	Cell density (cells/ml)	Purified E*r*-1 (μg/l)
0	180	ND[a]
24	580	6.2
48	1430	19.5
72	2810	50.4
96	3630	114

One-liter samples were removed daily from a 5-l culture of type I cells suspended with food, and analyzed for both cell density and amount of secreted pheromone E*r*-1.
[a]Not detectable.

culture filtrates onto reverse-phase cartridges, and then subjected to gelfiltration and anion-exchange chromatography on a Mono Q column. Marked variations in the amounts of purified protein distinguish one cell type from another, and they exactly reflect different degrees of amplification of the respective macronuclear coding genes (see Sect. 6, Pheromone Genes). From comparative analyses of maximal rates of pheromone secretion, it was found that type I cells on average secrete 20 pg of E*r*-1 per day, and that this amount is about twice that of E*r*-2 secreted from type II cells and ten times that of

Table 4. Maximum productions of different *E. raikovi* pheromones in standard cell cultures

Pheromone type	Secreted pheromones (μg/l)
E*r*-1	330[a]
E*r*-2	140[b]
E*r*-10	183[c]
E*r*-11	170[b]
E*r*-20	20[d]

[a]Data from Concetti et al. (1986).
[b]Data from Raffioni et al. (1987).
[c]Data from Raffioni et al. (1989).
[d]Data from Raffioni et al. (1992).

E*r*-20 secreted from type XX cells. The maximum pheromone productions estimated from cultures of different cell types suspended at concentrations of 5–10 × 10^3/ml are shown in Table 4.

In *E. octocarinatus*, hydroxyapatite is used for pheromone adsorption from culture filtrates (Weischer et al. 1985). Subsequently, pheromones G1 and G2 are brought to apparent homogeneity directly by gel filtration chromatography onto a Superose 12 column; the other two pheromones, G3 and G4, require preliminary chromatographic steps on anion-exchange DEAE-Sephadex A25 and Mono Q columns (Schulze Dieckhoff et al. 1987). From an overall comparison it appears that these pheromones are produced in much smaller amounts (from 40 to 800 times) than those of *E. raikovi* (G1, 0.5 μg/l; G3, 0.36 μg/l; G4, 0.38 μg/l).

Most purified pheromone preparations have been reported to be active from 10^{-11} to 10^{-13} M (Schulze Dieckhoff et al. 1987; Raffioni et al. 1989). This range, however, should be viewed with reserve, as it includes measures derived from the application of an unusual (but expedient) parameter, originally devised from studies on *Blepharisma* (Kubota et al. 1973; Miyake and Beyer 1974) and subsequently used also with other ciliates. It is based on the minimum amount of pheromone that is necessary to induce at least two cells of a given sample of assayed cells, ideally suspended without their pheromone, to form one (homotypic) mating pair. However, if the activity of the same pheromone is measured by the amount that is necessary for induction of at least 50% of the assayed cells to form mating pairs, then this activity would be in the range from 10^{-8} to 10^{-9} M (Ortenzi and Luporini 1995).

5 Pheromone Structure

Pheromones are synthesized in precursor forms, with the signal-peptide/prosegment/mature-protein alignment typical of precursors of secreted proteins. These pheromone precursors, denoted as prepropheromones, undergo at least

prepro

```
                    ▼
Er-1  (1-35) M N K L A I L A I I A M V L F S A N A F R F Q S R L R S N V E A K T G - -
Er-10 (1-37) M N K L A I L A I I A M V L F S A N A F R F Q S R I R S N V E A K T E T R
Er-2  (1-35) M N K L A I L A I I A M V L F S A N A F R L Q S R L R S N M E A - - S A R
Er-7  (1-35) M N K L A I L A I I A M V L F S A N A F R L Q S R L R L N M E A - - S A R
Er-11 (1-37) M N K L A I L A L A A M V L F S A V A F R L Q S M L R S N V E A Q T E T K
Er-20 (1-37) M N K L A I L A I I A M V L F S T N A F R L Q S K L R S N V E A Q T E T R
Er-21 (1-37) M N K L A I L A I I A M V L F S T N A F R F Q S K L R S N V E A Q T Q T R
```

secreted

```
Er-1  (36-75) D - - A C E Q A A I Q C V E S A C E S L C T E G E D R T G C Y M Y I Y S N - - C P P Y V
Er-10 (38-75) D - - L C E Q S A L Q C N E Q G C H N F C S - P E D K P G C L G M V W N P E L C P
Er-2  (36-75) D P M T C E Q A M A S C E H T M C - G Y C Q - G P L Y M T C I G I T T D P E - C G L P
Er-7  (36-75) D S M T C E Q A M A S C E H T M C - G Y C Q - G P L Y M T C I G I T T D P E - C A L P
Er-11 (38-76) D - - E C A N A A A Q C S I T L C N L Y C G - - P L I E I C E L T V M Q N - - C E P P F S
Er-20 (38-74) D - - I C D D A V A Q C S M T L C Q L - C Y - - - N T E I C E L S V I G S - - C Q P P F S
Er-21 (38-74) D - - I C D I A I A Q C S L T L C Q D - C V - - - N T P I C E L A V L G S - - C P P P W S
```

Fig. 1. Comparisons of amino acid sequences of seven pheromones of *E. raikovi*. *Gaps* were inserted to align the processing site of the prepro-domains and to maximize the identities of the secreted domains. The *arrowhead* indicates the putative processing site of the signal (pre) peptides

prepro

```
                          ▼
G3  (1-26)     M K A I F I I L A I L M V T Q A F K M T S K V N T K
G4  (1-42)     M K A I F I I L A I L M V T Q A F K M T S K V K S M

G3  (27-52)    L Q S Q I Q S K F Q S K N K L A S T F Q T S S Q L K
G4  (27-42)    N M S R N M S K N T S - - - - - - - - - - - T L G T K
```

secreted

```
G3  (53-77)    Y Y - - C W E E P Y T S S I T G C S T S L A C Y E A S
G4  (43-68)    Y T Y G C P Q T N T P T Q Q D - C Y D A M Y Y T F M A

G3  (78-104)   D C S V T G N D Q D K C N N V G Q N M I D K F F E L W
G4  (69-95)    M C D L Y P D P E H P M F P S Y D S C Q E E S D S A D

G3  (105-131)  G V C I N D Y E T C L Q Y V D R A W I H Y S D S E F C
G4  (96-102)   E F Y T N Q - - - - - - - - - - - - - - - - - - - - C

G3  (132-151)  G C T N P E Q E S A F R D A M D C L Q F
G4  (103-127)  G C G G Y - G M A A A H D Q V - C L L A L G V C I P E
```

Fig. 2. Comparisons of amino acid sequences of two pheromones of *E. octocarinatus*. *Gaps* and *arrowhead* as in Fig. 1

two proteolytic processing steps during exocytosis, i.e., the cotranslational removal of the signal peptide followed by the posttranslational activation of the pro-segment. In *E. raikovi*, they contain 70 amino acid residues, of which 19 form the pre-segment, 16–18 the pro-segment, and 37–40 the secreted pheromone (Miceli et al. 1989, 1991; Raffioni et al. 1992). In *E. octocarinatus*, they are twice as long as those of *E. raikovi*, and their three regions are formed by 16, 27–37, and 85–99 residues, respectively, with a total of 130–150 residues in the entire sequence (Meyer et al. 1992).

In Figs. 1 and 2, the seven pheromone sequences (E*r*-1, E*r*-2, E*r*-7, E*r*-10, E*r*-11, E*r*-20, and E*r*-21) so far determined as unique in *E. raikovi* (E*r*-3 and E*r*-9 have been found to be equivalent to E*r*-1 and E*r*-2, respectively) and the two (G1 and G2) of *E. octocarinatus* are aligned. From an overall comparison, it appears that the signal peptide is the most conserved region, not only within each species, but also between the two species. It appears to be proteolytically cleaved at a common Ala-Phe dipeptide and shows all the features described by Von Heijne (1987) as necessary for the activity of such entities: a positively charged amino terminus, a hydrophobic middle region, and a polar stretch immediately adjacent to the cleavage site. Significant variability begins with the pro-segment sequences, as also shown by their markedly different lengths. Nevertheless, there is one salient feature in common to all: there is a substantial repetition, from two (in *E. raikovi*) to several (in *E. octocarinatus*), of the motif Arg/Lys-Xaa-Xaa-Ser (with occasional substitutions of Lys with Met or Gln, and of Ser with Thr), that may be of functional importance for phosphorylation processes for protein kinase C (Kennelly and Krebs 1991). In addition, the putative cleavage site for producing the mature pheromone is characterized by a similar motif, which occurs also in other secreted proteins (Bond and Butler 1987): Arg/Lys-Asp in *E. raikovi*, but with the intriguing exception of Gly-Asp in the pro-E*r*-1, and Lys-Tyr in *E. octocarinatus*.

The sequences of the secreted pheromones represent the most variable regions. From interspecies comparisons, no salient feature in common is immediately evident, except an apparent conservation of three half-cystine residues. On the other hand, the sequences of the two *E. octocarinatus* pheromones show few similarities. Only 12 identities can be recognized after the insertion of a 20-gap segment into the carboxy-terminal half of G4, and one of these is represented by the amino-terminal Tyr residue and other six by six of the ten cysteines present in each sequence. More significant insights into the extent of similarity between these structures awaits relevant information from studies of recombinant G3, that has been recently expressed as a fusion protein in a bacterial vector (Brünen-Nieweler et al. 1994).

For *E. raikovi* pheromones, the comparison can be extended to seven unique amino acid sequences and, more importantly, to the structures of three of them, namely E*r*-1, E*r*-2, and E*r*-10, that have been determined by NMR spectroscopy (Luginbühl et al. 1994). All these sequences start with Asp at the amino terminus and contain six strictly conserved half-cystine residues, denoted as Cys I through Cys VI. As summarized in Table 5, most values of sequence identity are lower than 50%; yet there are sequences, such as those of E*r*-2 and E*r*-7, that are 95% identical. The various degrees of pheromone identity or divergence do not show any obvious direct relationship with the

Table 5. Comparison of amino acid sequences of different *E. raikovi* pheromones

	E*r*-1	E*r*-2	E*r*-7	E*r*-10	E*r*-11	E*r*-20	E*r*-21
E*r*-1	–	26	26	41	30	25	27
E*r*-2		–	95	30	34	27	27
E*r*-7			–	30	34	27	27
E*r*-10				–	26	24	26
E*r*-11					–	59	51
E*r*-20						–	73
E*r*-21							–

Values are given in percentages and calculated on the basis of the alignments shown in Fig. 1, with gaps counted as non-identities.

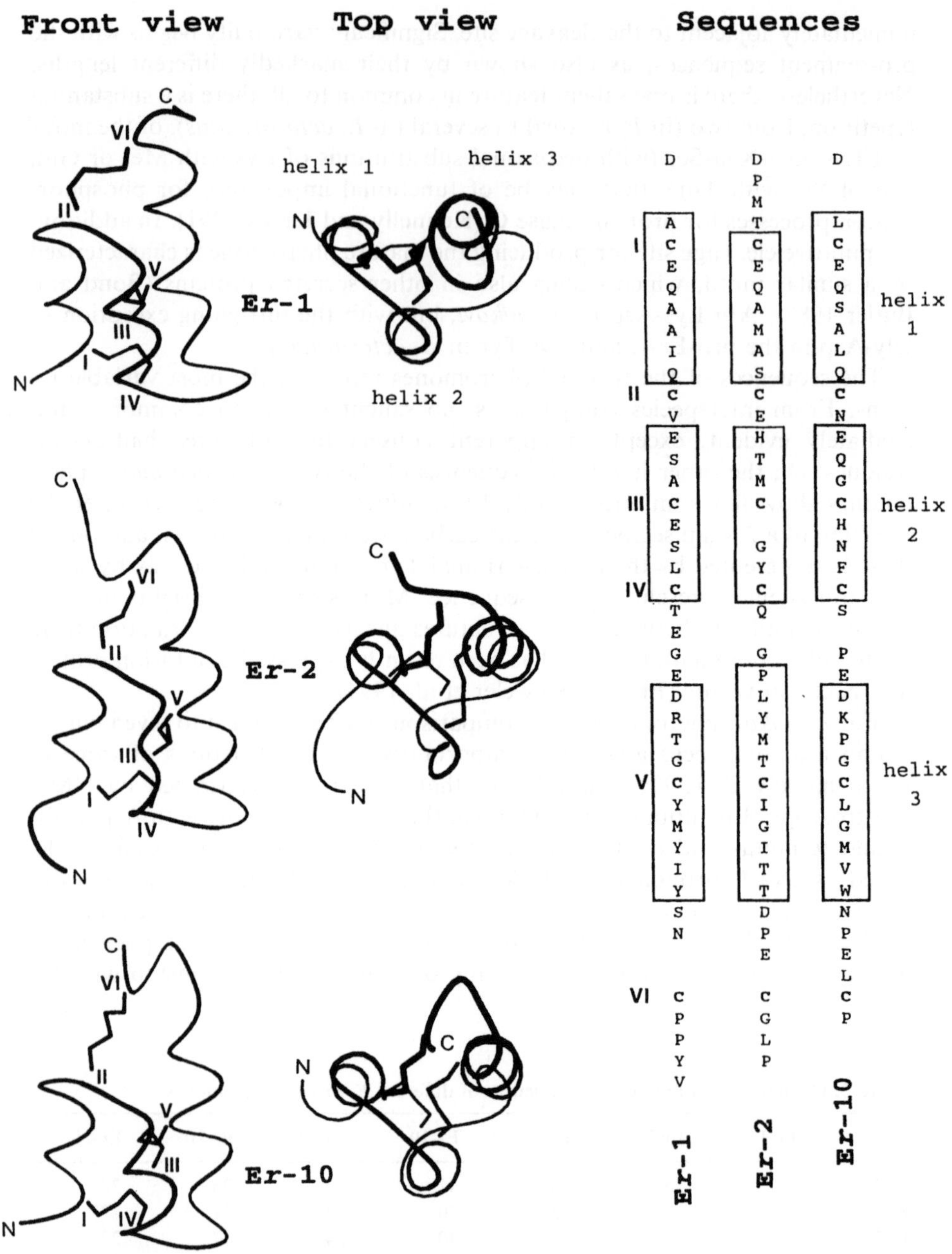

Fig. 3. Schematic comparison of structural features between E*r*-1, E*r*-2, and E*r*-10. On the *left* backbone ribbons and disulfide bridges from NMR models of the three pheromones; *thicker lines* indicate regions of the molecules that are closer to the viewer; disulfide bridges are represented by *notched lines*. On the *right* pheromone sequences presented *from top to bottom*; the *boxes* indicate the locations of the helices. The half cystine residues are marked by progressive *Roman numerals*

spectra of cell-mating reactions, in particular with the possible pheromone separation into distinct subfamilies (see above Table 1). Percent values of pheromone identities within each subfamily may be low, as in interfamily comparisons.

Regardless of the deep divergences in their primary structure, *E. raikovi* pheromones all apparently conform to a common architecture, typical of small, highly stable, and disulfide-rich proteins. This was first evident from the determination of the disulfide bond pairing in E*r*-1 and E*r*-2, as this pairing (Cys I-Cys IV, Cys II-Cys VI, and Cys III-Cys V) is equivalent in both sequences, which are only 26% identical (Stewart et al. 1992). As illustrated in Fig. 3, the E*r*-1, E*r*-2, and E*r*-10 architectures – which have been analyzed and compared in great detail by Luginbühl et al. (1994) – are all based on a bundle of three helices with nearly parallel axes and an up-down-up topological arrangement. The second (central) helix is connected to the first and the third helix by two loops of two and two-four residues, respectively. These interhelix connections are strongly stabilized by two disulfide bonds placed in close proximity with each other; one (Cys I-Cys IV) lies between the first and the second helix, the other (Cys III-Cys V) between the second and the third one. Further tightness of the molecular structure is finally achieved by means of the third disulfide bond (Cys II-Cys VI) that links the carboxy-terminal region to the connection between the first and the second helix. Overall, the pheromone shape resembles a pyramid with a triangular basis indented by a deep cleft, and three faces formed by two helices each; one face is mainly hydrophobic, another charged and aromatic, and the third one less characterized.

Although pheromones mimic each other in their overall architecture, there are significant local variations of their structures. Thus they also possess clearcut "personal" features of potential functional importance for conferring specificity to their receptor-binding reactions. Main distinctive variations occur in the organization of the second helix (that changes in E*r*-2 from α- to 3_{10}-helix to accommodate one deletion), in the shape of the loop connecting this helix to the third one, and the orientation and length of the terminal segment extending from the end of the third helix to the carboxy-terminus. This segment, in particular, looks like a distinctive hallmark of each pheromone. It is six residues long in E*r*-10 and seven in E*r*-1 and E*r*-2, and its spatial arrangement appears to be largely conditioned by the position assumed therein by two Pro residues and Cys VI, which is involved in the disulfide pairing with the relatively distant Cys II. This pairing can more readily be formed in E*r*-1, in which only two residues separate Cys VI from the end of the third helix, than in E*r*-2 and E*r*-10, in which this separation involves three or four residues, respectively. As a consequence, the terminal segment of E*r*-1 curls up over the top of the third helix, whereas those of E*r*-2 and E*r*-10 fold back and dangle from the helix bundle as whips of different lengths.

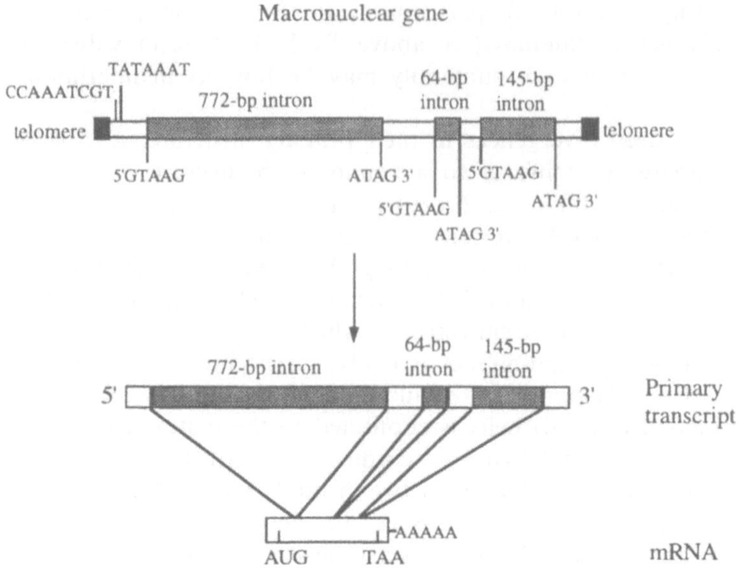

Fig. 4. Schematic representation of the structural organization of the macronuclear gene and encoded mRNA for pheromone G4 of *E. octocarinatus*. CAAT and TATA box-like elements and splice-junction sequences are here added to the scheme of the macronuclear gene. (After Meyer et al. 1992)

6 Pheromone Genes

Pheromones are specified by genes which conform structurally with all the other genes so far characterized in the subchromosomal genome of the macronucleus of hypotrich ciliates. This genome derives from the chromosomal genome of the micronucleus through complex and extensive processes of DNA diminution and rearrangement coincident with the cell reorganization following a sexual process, and contains only linear DNA molecules acting as "free" genes as regards the control of replication and transcription (Klobutcher and Prescott 1986; Jahn 1991). Each gene is usually represented by thousands of copies and consists of a central coding region flanked by two, usually shorter, noncoding regions terminating with telomeres characterized by repeats of the octanucleotide 5′CCCCAAAA3′.

In *E. octocarinatus*, the genes encoding pheromones G3 and G4 have been characterized as linear DNA molecules of about 1700 bp (Meyer et al. 1991, 1992), basically organized as illustrated in Fig. 4. In contrast to other known *Euplotes* genes, which usually lack common eukaryotic control signals for the regulation of transcription (Ghosh et al. 1994), the noncoding region at the 5′ end of the G4 gene of *E. octocarinatus* has been shown to contain potentially regulatory CAAT and TATA box-like elements (Meyer et al. 1992). In the coding region there are three introns, one of which, 772 bp long, interrupts the sequence corresponding to the pheromone signal peptide, and the other two, of 145 and 64 bp, interrupt that which corresponds to the secreted pheromone. Although all three introns similarly carry typical eukaryotic 5′ and 3′ splice junctions, and apparently conserve their relative positions in different genes (Brünen-Nieweler et al. 1991; Meyer et al. 1992), a remarkable feature distinguishes them. While the G-C content of the two short introns is relatively low, as is common in introns, that of the longer one is unusually high, as in exons. As this intron also contains potential reading frames, it might be capable of coding functions.

The study of these pheromone genes has also contributed to our knowledge of the eccentric use of the genetic code by ciliates, discovered in *Paramecium* and *Tetrahymena*, which specify glutamine by means of the codons UAA and UAG (as a review, Martindale 1989). In both the G3 and G4 genes, the UGA codon – a stop codon – is used, in addition to the UGU and UGC codons, to specify cysteine (Meyer et al. 1991).

In *E. raikovi*, the pheromone genes are only about 1100 bp long (Miceli et al. 1989, 1991). No canonical eukaryotic regulatory sequence is present in their noncoding regions. Nevertheless, it has been suggested that their expression might be modulated by a mechanism of differential gene amplification (La Terza et al. 1995). The genes encoding pheromones E*r*-1, E*r*-2, and E*r*-10, are represented, in homozygous as well as in heterozygous cells, by clearly distinct numbers of copies ($2.5–2.9 \times 10^4$, $0.9–1.2 \times 10^4$, and $1.6–1.8 \times 10^4$, respectively), and the ratios between the different degrees of amplification (E*r*-1/E*r*-2, 2.4;

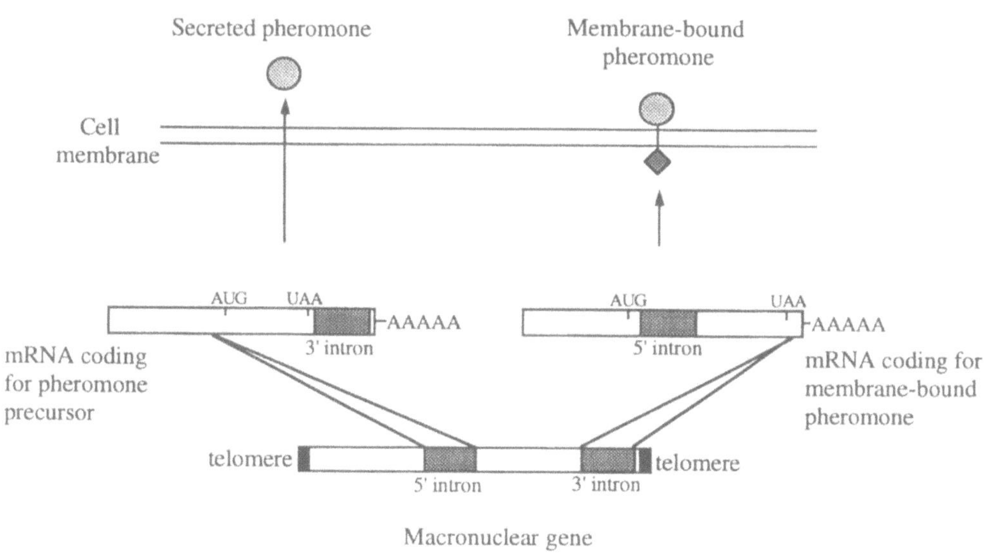

Fig. 5. Schematic representation of the production of secreted and membrane-bound E*r*-1 pheromone isoforms from a single macronuclear gene by a process of alternative splicing in type I cells of *E. raikovi*

```
                                   -1 1
-19       5' TAGCCTATAATGTCATAGA ATG GGA TGT ACA TCA GAT CTT TGT CAC ATG GCA TCA GTG  39
                                 Met Gly Cys Thr Ser Asp Leu Cys His Met Ala Ser Val
                                  1                                  10

 40  TTA AGT AAA TAT CAG AGC CTA TTT CTA TAT CTG TGC CGT TCA AAT AAC TGT GTT GGT CCG TTA 102
     Leu Ser Lys Tyr Gln Ser Leu Phe Leu Tyr Leu Cys Arg Ser Asn Asn Cys Val Gly Pro Leu
                             20                                      30

103  AAC TCT ATC AAT AGG TTC TTC TCT GGC ACA GTA CAT GTC ATA AGT TTT AGT ATC AAT TTT AGA 165
     Asn Ser Ile Asn Arg Phe Phe Ser Gly Thr Val His Val Ile Ser Phe Ser Ile Asn Phe Arg
                         40                                  50

166  ATG AAC AAA CTA GCA ATT CTC GCT ATC ATC GCT ATG GTA CTC TTC AGC GCC AAC GCC TTC AGA 228
     Met Asn Lys Leu Ala Ile Leu Ala Ile Ile Ala Met Val Leu Phe Ser Ala Asn Ala Phe Arg
                     60                                  70

229  TTC CAA AGC AGA TTG AGA TCA AAT GTA GAA GCT AAG ACA GGA GAT GCT TGT GAG CAA GCT GCA 291
     Phe Gln Ser Arg Leu Arg Ser Asn Val Glu Ala Lys Thr Gly Asp Ala Cys Glu Gln Ala Ala
                 80                                      90

292  ATC CAG TGT GTT GAG TCA GCA TGT GAA AGT CTT TGT ACA GAA GGT GAA GAT AGA ACT GGC TGC 354
     Ile Gln Cys Val Glu Ser Ala Cys Glu Ser Leu Cys Thr Glu Gly Glu Asp Arg Thr Gly Cys
             100                                     110

355  TAT ATG TAC ATC TAT TCT AAC TGC CCA CCT TAT GTC TAA ATCTAACTACTTTAGAATCAAACCAAATTCG 424
     Tyr Met Tyr Ile Tyr Ser Asn Cys Pro Pro Tyr Val Stop
         120                                     130

425  GACAAGTTGTAGTTAACACTTACTCTGGAGTTACTTGACAAGAACTAGGATAACTTTATTGAACAAAGTTCCATCT  3'     500
```

Fig. 6. Complete coding sequence and part of the flanking noncoding regions of the macronuclear gene for pheromone Er-1 of *E. raikovi.* The two initiation translation ATG codons for the membrane-bound and secreted pheromone isoforms are *underlined* at nucleotide positions 1 and 166, respectively. The *framed sequences* close to the 5′ and 3′ end correspond to the 5′- and 3′- introns, respectively, that are alternatively removed in the mRNAs

E*r*-1/E*r*-10, 1.6; E*r*-10/E*r*-2, 1.6) are closely comparable with the ratios of the amounts of the respective mRNAs synthesized and pheromones secreted (La Terza et al. 1995).

Like the G3 and G4 pheromone genes of *E. octocarinatus*, *E. raikovi* pheromone genes also contain intron-like elements which, at least in the case of the E*r*-1 gene, show unusual splice junctions (Miceli et al. 1992). The process of removal of these elements was first determined by analyses of E*r*-1 cDNA clones prepared from type I cells. It has now been confirmed by a similar study carried out on E*r*-2 cDNA clones prepared from type II cells (Miceli et al. 1994 and unpubl. data), thus suggesting that it may be of general occurrence in every *E. raikovi* cell type. As schematically illustrated in Fig. 5, in the E*r*-1 gene there are two intron-like elements (that were originally denoted as IVS, i.e., intervening sequences), one of 89 bp located in the 5′ half of the gene and the other of 78 bp in the 3′ half. They (hereafter, for the sake of simplicity, referred to as the 5′- and 3′-introns) are alternatively removed from the primary transcript, with the production of two distinct mRNAs. The mRNA without the 5′-intron encodes the 75-residue preproE*r*-1, that is processed for E*r*-1 secretion. The mRNA with the 5′-intron has a longer reading frame and specifies a protein of 130 amino acid residues, yielding a calculated molecular mass of about 14 kDa. This protein, the sequence of which is reported in Fig. 6, is retained in the cell membranes, as a "type II" transmembrane protein (Wickner and Lodish 1985), in such a way that its 75 residues at the carboxy-terminus, which are identical with those forming the prepro-E*r*-1, represent its extracellular and transmembrane domains. The other 55 residues at the amino-terminus are new and represent its cytoplasmic domain, showing at least two sequence motifs, Lys-Tyr-Gln-Ser and Arg-Phe-Phe-Ser, that might be sites of modification by protein kinases (Kennelly and Krebs 1991). It is this membrane-anchored, cell type-specific pheromone isoform that present experimental evidence (described, in part, in the next section) suggests represents the fundamental binding unit of the autocrine pheromone receptors.

7 Pheromone Receptors

Evidence that cells produce autocrine pheromone receptors for the same pheromones that they secrete was first obtained in *E. raikovi* by studies of binding of radioiodinated preparations of E*r*-1 and E*r*-2 to type I and type II cells, respectively (Ortenzi et al. 1990; Luporini et al. 1992). As shown in Fig. 7, this binding was typically dose-responsive and saturable, and analysis of the binding data on Scatchard plots usually revealed the best fit to be a straight line, implying that homozygous cells secreting one pheromone carry homogeneous populations of pheromone-binding sites. These sites showed values of the dissociation constant (K_d) of the order of 10^{-9} M in both type I and type II cells, but their average numbers per cell were two to three times higher in the former than in the latter, i.e., 29–50 × 10^6 vs. 16 × 10^6, respectively. These numbers seem to be unusually high in absolute terms; nevertheless, they reflect normal receptor densities on cells of more advanced organisms. They are equivalent to a maximal distribution of about 1700 sites per μm^2 of the *E. raikovi* cell surface, since this can be estimated to measure about 30 000 μm^2 (including the extensions of its hundreds of cilia), i.e., much larger than the surface of a mammalian cell of average size.

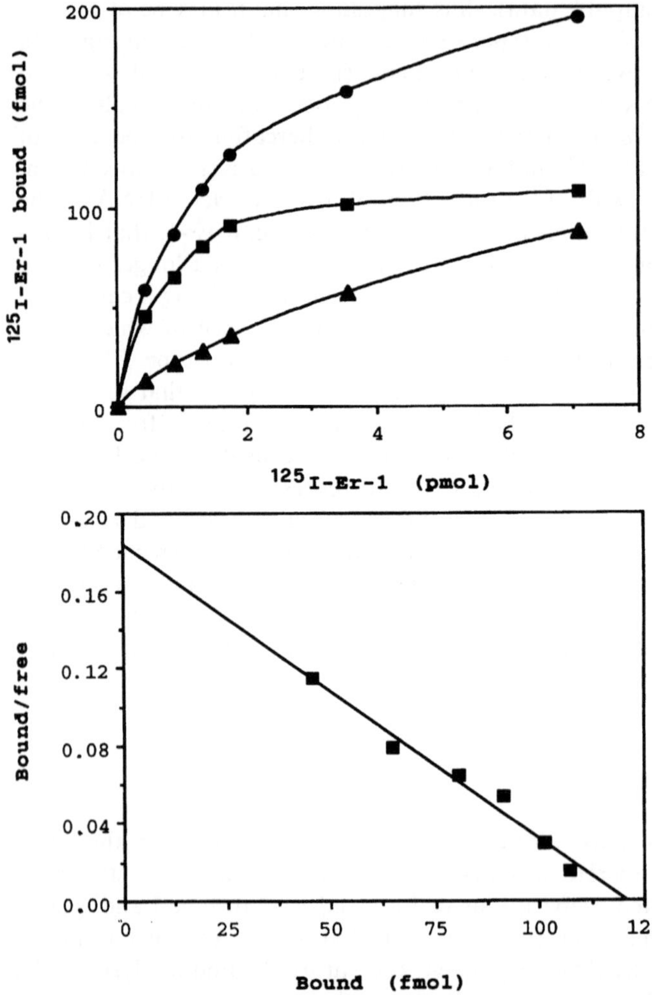

Fig. 7. Concentration dependence and Scatchard analysis of ^{125}I-E*r*-1 binding to type I cells in stationary phase. Cells (2.5 x 10^3) were incubated for 1 h at 24 °C with increasing concentrations of ^{125}I-E*r*-1, and final volumes of 250 μl were adjusted with buffer, as reported by Ortenzi et al. (1990). Values are from a single experiment and represent total binding (●), specific binding (■), and nonspecific binding (▲)

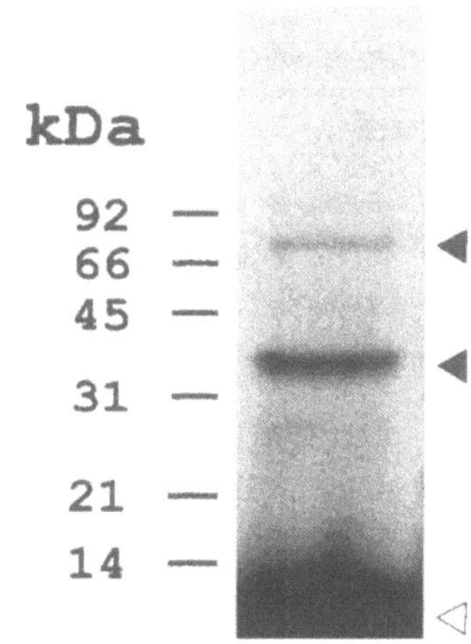

Fig. 8. SDS-PAGE and autoradiographic analysis of solubilized cell membrane preparations cross-linked to ^{125}I-E*r*-1. Membranes were prepared and cross-linked as described by Ortenzi et al. (1990), and the specific activity of the radiolabeled pheromone was 100 μCi/μg. The *black arrowheads* indicate the position of the pheromone-receptor complexes; the *white arrowhead* the position of the unbound radioligand

Analyses of chemical cross-linking of ^{125}I-E*r*-1 and ^{125}I-E*r*-2 to preparations of solubilized cell membranes (Ortenzi et al. 1990; Luporini et al. 1992), and of pheromone-binding activity to proteins purified by ligand affinity chromatography (unpubl. results) have provided evidence suggesting that these pheromone-binding sites consist of a 14-kDa membrane-bound pheromone isoform controlled by the alternative splicing mechanism described above. It seems likely that this isoform undergoes oligomerization in pheromone binding reactions. As shown in Fig. 8, cross-linking in fact reveals the formation of one major band at M_r of 38 kDa, that is the size expected from the association of two complexes between a pheromone and its membrane-bound isoform; and, occasionally, when radiolabeled pheromone preparations with a very high specific activity are used, a minor band at M_r of 76 kDa becomes evident, suggesting the formation of tetrameric complexes. Similarly, from affinity chromatography it is possible to identify one principal pheromone binding protein of about 14 kDa that usually coexists, in markedly variable quantitative ratios, with at least another species of about 28 kDa, probably representing a homodimeric association of the 14-kDa protein.

Important insights into the possible pheromone interactions with the extracellular portion of its putative receptor (represented by its membrane-bound isoform) have recently been derived from the definition of the crystal structure of pheromone E*r*-1 at 1.6 Å resolution (Weiss et al. 1995), as such interactions may be effectively represented by the pheromone interactions that occur in the crystal. Two distinct modes of association of the pheromone molecules are observed. Both involve helix-helix interactions of neighboring monomers and can occur simultaneously. The relative dimerization energies are, in fact, similar, and the dimerization interfaces of the molecules do not overlap. One association is due to the formation of symmetrical dimers, which are basically four-helix bundles comprised of the first and second helices from both

Table 6. Comparison of dissociation constants (K_d) obtained from the binding of ^{125}I-E*r*-1 and ^{125}I-E*r*-2 to different cell types of *E. raikovi*

Cell type	$K_d \times 10^9$ M (± SE)	
	^{125}I-E*r*-1	^{125}I-E*r*-2
I	6.30 ± 0.2	7.47 ± 0.3
II	6.33 ± 0.1	6.45 ± 0.4
X	9.72 ± 0.7	8.83 ± 0.4

monomers; the other to the formation of asymmetric dimers resulting mainly from stacking the third helix in an antiparallel fashion.

8 Competitive Pheromone Receptor-Binding Reactions

Assays of the capability of cells exposed to a "foreign" (nonself) pheromone to form (homotypic) mating pairs reveal that the extent of the cell-mating response to a nonself pheromone depends not only on the concentration of this pheromone that has been added to the cells, but also on the concentration of

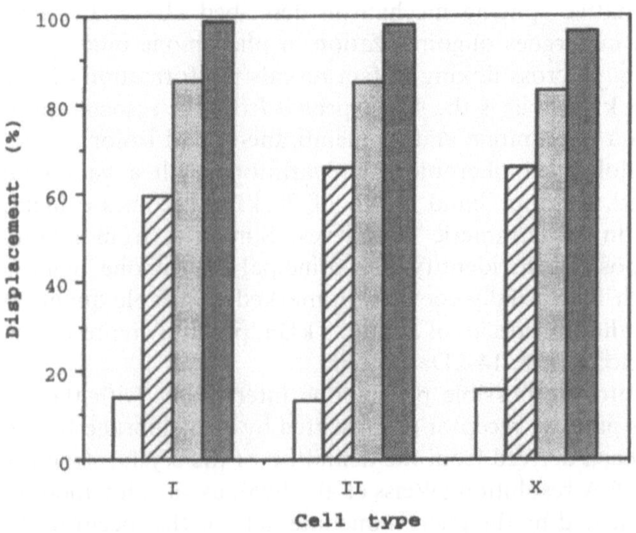

Fig. 9. Inhibition of ^{125}I-E*r*-1 binding to cells of types I, II, and X by unlabeled E*r*-1. Cells (2.5 x 10^3 in 250 µl of seawater) were incubated with 150 ng/ml of ^{125}I-E*r*-1 in the presence of increasing concentrations of the unlabeled E*r*-1, at 4 °C. After 1 h of incubation, specifically bound ^{125}I-E*r*-1 was measured. For each cell type, the concentrations of unlabeled E*r*-1 relative to that of ^{125}I-E*r*-1 were equimolar (*white columns*), 5-fold molar (*hatched columns*), 25-fold molar (*light gray columns*), 100-fold molar (*dark gray columns*). Values represent means of duplicate measurements

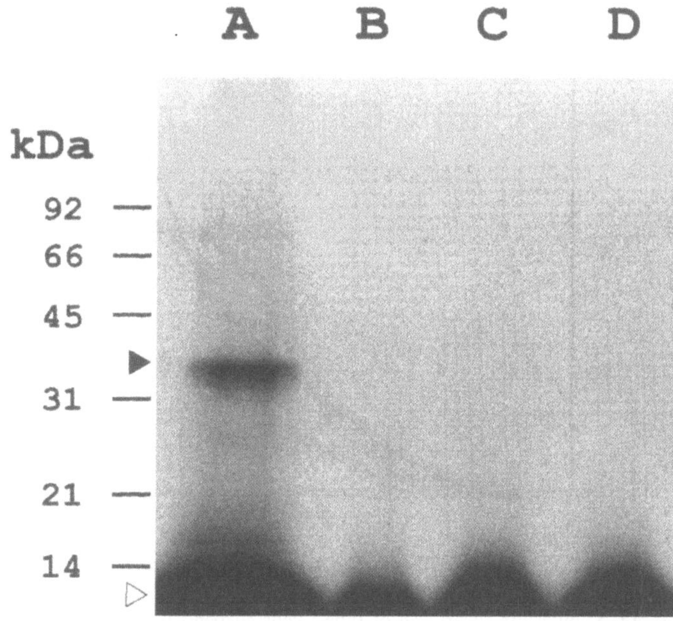

Fig. 10. SDS-PAGE and autoradiographic analysis of solubilized cell membrane preparations cross-linked to ^{125}I-E*r*-1. Lane *A*, membrane preparations (350 μg) from cells of type I cross-linked to ^{125}I-E*r*-1 (100 ng). Lanes *B*, *C*, and *D*, as in lane *A* but with added excess (10 μg) of unlabeled pheromones E*r*-1, E*r*-2, and E*r*-10, respectively. Membranes were prepared and cross-linked as described by Ortenzi et al. (1990). The specific activity of the radiolabeled pheromone was 2.9 μCi/μg. The *black arrowhead* indicates the position of the major pheromone-receptor complex; the *white arrowhead* indicates the position of the unbound radioligand

the cell's own (self) pheromone that is already present in the medium (Luporini and Miceli 1986). Moreover, addition of increasing concentrations of self or nonself pheromones to loosely paired cells clearly favors or prevents, respectively, their subsequent separation. Results drawn from these observations, indicating that different (self and nonself) pheromones can mutually compete for binding to each other's cell receptors and elicit a shift from the vegetative to the mating stage and vice versa, have now been substantiated by results from in vitro pheromone-binding experiments (Ortenzi and Luporini 1995). As reported in Table 6, cell-binding reactions of pheromones which are members of the same family (and hence share a common architecture), such as E*r*-1 and E*r*-2, occur with similar, but not identical, K_d values of the order of 10^{-9} M. As shown in Figs. 9 and 10, a minimum excess of one pheromone can effectively inhibit the receptor-binding reactions of another one.

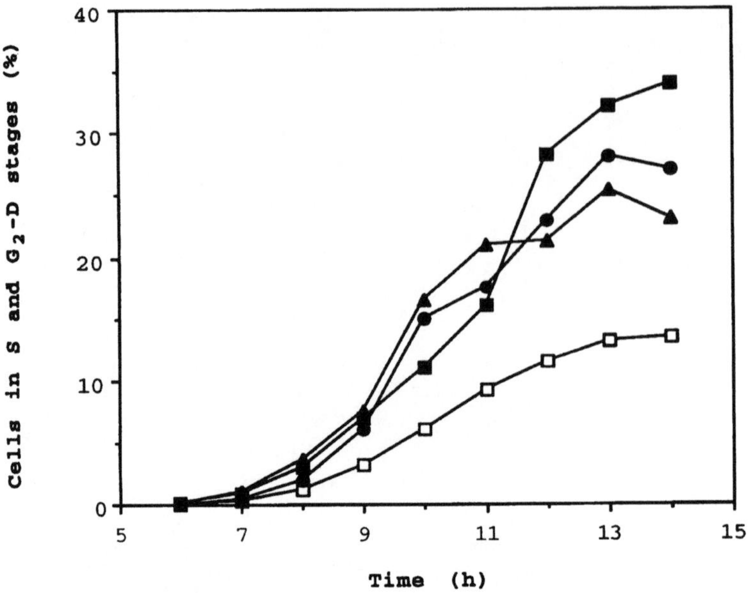

Fig. 11. Stimulation of growth of type I cells suspended in medium containing E*r*-1 pheromone. Cells were partially synchronized by starvation and time-limited refeeding to enter a new division cycle and then suspended in: fresh seawater (□); sterilized cell supernatant (▲); 34 nM E*r*-1 (●); and 340 nM E*r*-1 (■). At the indicated intervals, cell samples were removed from the suspensions and stained with Feulgen reagent. Cells in S and G_2-D stages were distinguished by the presence of replication bands in the macronuclei. Percentages were calculated by observations on at least 300 cells from each of three individual samples

9 Pheromones as Growth Factors

The finding that pheromones bind to specific receptors borne by the cells from which they are constitutively secreted throughout the cell life cycle was an obvious indication that these molecules could actually function as autocrine growth factors (Ortenzi et al. 1990; Luporini et al. 1992). Confirmation of this indication was derived from the observation that *E. raikovi* pheromones not only bear some intriguing structural similarities to mammalian IL-2 (Ortenzi et al. 1990; Luporini et al. 1994), but also that they can cross-react with some specific membrane components of T lymphocytes which are IL-2-dependent for survival and proliferation (Vallesi et al., ms in prep.).

Direct evidence for such a function of *E. raikovi* pheromones as autocrine growth factors has been recently derived from comparative analyses of rates of thymidine incorporation and cell-cycle progression in cells resuspended with or without their pheromone after having been removed from starved cultures at the resting (or G_0) stage, subjected to repeated washing to remove secreted pheromone, and allowed by a time-limited addition of food to move, in partial synchrony, into a new division cycle (Vallesi et al. 1995). As shown in Fig. 11, cells interacting with their pheromone enter the S-stage of the cell cycle, and then start dividing, in higher percentages, and earlier, than cells suspended in medium initially lacking pheromone.

10 Concluding Remarks

New insights into the biology and activity of ciliate pheromones have been derived from studies on *E. raikovi* and *E. octocarinatus*. Unlike the intriguing but exceptional case of *B. japonicum* "gamones", and in full accord with their genetic mechanisms, these *Euplotes* cell markers form structurally homologous protein families. Each family member contains common as well as unique structural features, that make it similar to, and at the same time distinct from, all the other family members. In addition, evidence has been presented that cells carry specific receptors for the same pheromones as those which they themselves constitutively secrete from the very beginning of their life cycle. Taken together, these findings provide a rationale for a better understanding of how pheromones can competitively bind to their cell receptors in both autocrine and paracrine fashions, and consequently elicit cell responses affecting both vegetative reproduction, or growth, and mating-pair formation. In this chapter, attention has been restricted to a presentation of *Euplotes* pheromones. It should not be forgotten, however, that a characterization of pheromones of other ciliates is essential for the definition of the functional and evolutionary relationships between these signal molecules and similar ones in more advanced multicellular organisms. In this perspective, structural information on the allelic pheromones of *Dileptus anser* (Parfenova et al. 1989), which have been reported as being able to induce both cell mating and reproduction (Tavrovskaja 1979), would be particularly instructive.

Acknowledgments. It is a pleasure to acknowledge the understanding collaboration of Prof. G.H. Beale in the preparation of this chapter, and the financial support provided by the Italian Ministero dell' Università e della Ricerca Scientifica e Tecnologica and CNR.

References

Akada R (1985) Mating types and mating-inducing factors (gamones) in the ciliate *Euplotes patella* syngen 2. Genet Res Camb 46: 135–132

Akada R (1986) Partial characterization of the mating-inducing factors (gamones) produced by mating type VI cells in the ciliate *Euplotes patella* syngen 2. J Exp Zool 237: 287–290

Beale GH (1954) The genetics of *Paramecium aurelia*. Cambridge Univ Press, London

Beale GH (1990) Self and nonself recognition in the ciliate protozoan *Euplotes*. Trends Genet 6: 137–139

Bond JS, Butler OE (1987) Intracellular proteases. Annu Rev Biochem 56: 333–364

Brünen-Nieweler C, Schmidt HJ, Heckmann K (1991) Two introns in the pheromone 3-encoding gene of *Euplotes octocarinatus*. Gene 109: 233–237

Brünen-Nieweler C, Meyer F, Heckmann K (1994) Expression of the pheromone 3-encoding gene of *Euplotes octocarinatus* using a novel bacterial secretion vector. Gene 150: 187–192

Caprette CL, Gates MA (1994) Quantitative analyses of interbreeding in populations of *vannus*-morphotype *Euplotes*, with special attention to the nominal species *E. vannus* and *E. crassus*. J Euk Microbiol 41: 316–324

Concetti A, Raffioni S, Miceli C, Barra D, Luporini P (1986) Purification to apparent homogeneity of the mating pheromone of *mat*-1 homozygous *Euplotes raikovi*. J Biol Chem 261: 10582–10586

Dini F, Nyberg D (1993) Sex in ciliates. In: Jones JG (ed) Advances in microbial ecology. Plenum Press, New York, 13: 85–153

Ghosh S, Jarakzewsky JW, Klobutcher LA, Jahn CL (1994) Characterization of transcription initiation, translation initiation, and poly(A) addition sites in gene-sized macronuclear DNA molecules of *Euplotes*. Nucl Acid Res 22: 214–221

Hartmann M, Schartau O (1939) Untersuchungen über die Befruchtungsstoffe der Seeigel. I. Biol Zentralbl 59: 571–587

Heckmann K, Kuhlmann HW (1986) Mating types and mating inducing substances in *Euplotes octocarinatus*. J Exp Zool 237: 87–96

Hiwatashi K (1988) Sexual recognition in *Paramecium*. In: Chapman GP (ed) Eukariote cell recognition – concepts and model systems. Cambridge Univ Press, London, pp 77–91

Jaenicke L (1984) Biological activity of blepharismone derivatives. An approach towards biogenesis of the hormone. In: Schlossberger HG, Kochen W, Linzen B, Steinhart H (eds) Progress in tryptophan and serotonin research. De Gruyter, Berlin, pp 815–826

Jahn CL (1991) The nuclear genome of hypotrichous ciliates: maintaining the maximum and the minimum of information. J Protozool 38: 252–258

Kennelly P, Krebs EG (1991) Consensus sequences as substrate specificity determinants for protein kinases and protein phosphatases. J Biol Chem 266: 1555–1558

Kimball RF (1942) The nature and inheritance of mating types in *Euplotes patella*. Genetics 27: 269–285

Kitamura A (1988) Mating-type substances. In: Gortz HD (ed) *Paramecium*. Springer, Berlin Heidelberg New York, pp 85–94

Klobutcher LA, Prescott DM (1986) The special case of the hypotrichs. In: Gall JG (ed) The molecular biology of ciliated protozoa. Academic Press, Orlando, pp 111–154

Kubota T, Tokoroyama T, Tsukuda Y, Koiama H, Miyake A (1973) Isolation and structure determination of blepharismin, a conjugation initiating gamone in the ciliate *Blepharisma*. Science 179: 400–402

Kuhlmann HW, Heckmann K (1989) Adolescence in *Euplotes octocarinatus*. J Exp Zool 251: 316–328

Kusch J, Heckmann K (1988) Gamones are secreted in *Euplotes octocarinatus* via the cortical ampules. Eur J Protistol 23: 273–278

La Terza A, Miceli C, Luporini P (1995) Differential amplification of pheromone genes of the ciliate *Euplotes raikovi*. Dev Genet 17: 272–279

Luginbühl P, Ottiger M, Mronga S, Wüthrich K (1994) Comparison of the NMR structures of the homologous pheromones E*r*-1, E*r*-10 and E*r*-2 from *Euplotes raikovi*. Protein Sci 3: 1537–1547

Luporini P, Miceli C (1986) Mating pheromones. In: Gall JG (ed) The molecular biology of ciliated protozoa. Academic Press, New York, pp 263–299

Luporini P, Raffioni S, Concetti A, Miceli C (1986) The ciliate *Euplotes raikovi* heterozygous at the *mat* genetic locus coreleases two individual species of mating pheromones: genetic and biochemical evidence. Proc Natl Acad Sci USA 83: 2889–2893

Luporini P, Miceli C, Ortenzi C, Vallesi A (1992) Developmental analysis of the cell recognition mechanism in the ciliate *Euplotes raikovi*. Dev Genet 13: 9–15

Luporini P, Vallesi A, Miceli C, Bradshaw RA (1994) Ciliate pheromones as early growth factors and cytokines. In: Beck G, Cooper EL, Habicht GS, Marchalonis JJ (eds) Primordial immunity: foundations for the vertebrate immune system. Ann N Y Acad Sci 712: 195–205

Martindale DW (1989) Codon usage in *Tetrahymena* and other ciliates. J Protozool 36: 29–34

Meyer F, Schmidt HJ, Plumber E, Hasilik A, Mersmann G, Meyer H, Engstrom A, Heckmann K (1991) UGA is translated as cysteine in pheromone 3 of *Euplotes octocarinatus*. Proc Natl Acad Sci USA 88: 3758–3761

Meyer F, Schmidt HJ, Heckmann K (1992) Pheromone 4 gene of *Euplotes octocarinatus*. Dev Genet 13: 16–25

Miceli C, La Terza A, Melli M (1989) Isolation and structural characterization of cDNA clones encoding the mating pheromone E*r*-1 secreted by the ciliate *Euplotes raikovi*. Proc Natl Acad Sci USA 86: 3016–3020

Miceli C, La Terza A, Bradshaw RA, Luporini P (1991) Structural characterization of mating pheromone precursors of the ciliate protozoan *Euplotes raikovi*. Eur J Biochem 202: 759–764

Miceli C, La Terza A, Bradshaw RA, Luporini P (1992) Identification and structural characterization of a cDNA clone encoding a membrane-bound form of the polypeptide pheromone E*r*-1 in the ciliate protozoan *Euplotes raikovi*. Proc Natl Acad Sci USA 89: 1988–1992

Miceli C, Di Giuseppe G, Luporini P (1994) The macronuclear gene encoding pheromone E*r*-2 of *Euplotes raikovi*. J Euk Microbiol 41: 6A

Miyake A (1978) Cell communication, cell union and initiation of meiosis in ciliate conjugation. Curr Top Dev Biol 12: 37–82

Miyake A (1981a) Cell interactions by gamones in *Blepharisma*. In: O'Day DH, Horgen P (eds) Sexual interactions in eucaryotic microbes. Academic Press, New York, 4: 95–129

Miyake A (1981b) Physiology and biochemistry of conjugation in ciliates. In: Levandowsky M, Hutner SH (eds) Biochemistry and physiology of protozoa. Academic Press, New York, 4: 125–198

Miyake A (1983) Gamones in ciliates. J Protozool 30: 43A

Miyake A, Beyer J (1974) Blepharmone: a conjugation-inducing glycoprotein in the ciliate *Blepharisma*. Science 185: 621–623

Nanney DL (1980) Experimental ciliatology. Wiley, New York

Ortenzi C, Luporini P (1995) Competition among homologous polypeptide pheromones of the ciliate *Euplotes raikovi* for binding to each other's cell receptors. J Euk Microbiol 42: 242–248

Ortenzi C, Miceli C, Bradshaw RA, Luporini P (1990) Identification and initial characterization of an autocrine pheromone receptor in the protozoan ciliate *Euplotes raikovi*. J Biol 111: 607–614

Parfenova EV, Afon'kin SY, Yudin AL, Etingof RN (1989) Characterization and partial purification of mating pheromone excreted by mating type II cells of the ciliate *Dileptus anser*. Acta Protozool 28: 11–21

Raffioni S, Miceli C, Concetti A, Barra D, Luporini P (1987) Purification and characterization of new mating pheromones of the ciliate *Euplotes raikovi*. Exp Cell Res 172: 417–424

Raffioni S, Luporini P, Bradshaw R (1989) Purification, characterization, and amino acid sequence of the mating pheromone E*r*-10 of the ciliate *Euplotes raikovi*. Biochemistry 28: 5250–5256

Raffioni S, Miceli C, Vallesi A, Chowdhury SK, Chait BT, Luporini P, Bradshaw RA (1992) Primary structure of *Euplotes raikovi* pheromones: comparison of five sequences of pheromones with variable mating interactions. Proc Natl Acad Sci USA 89: 2071–2075

Schlegel M (1991) Protist evolution and phylogeny as discerned from small subunit ribosomal RNA sequence comparisons. Eur J Protistol 27: 207–219

Schulze Dieckhoff H, Freiburg M, Heckmann K (1987) The isolation of gamones 3 and 4 of *Euplotes octocarinatus*. Eur J Biochem 168: 89–94

Sonneborn TM (1937) Sex, sex inheritance and sex determination in *Paramecium aurelia*. Proc Natl Acad Sci USA 23: 378–385

Stewart AE, Raffioni R, Chaudhary T, Chait BT, Luporini P, Bradshaw RA (1992) The disulfide bond pairing of the pheromones E*r*-1 and E*r*-2 of the ciliate protozoan *Euplotes raikovi*. Protein Sci 1: 777–785

Tavrovskaja MV (1979) Intraspecific intercellular interaction in the ciliate *Dileptus anser*. J Protozool 26: 35A–36A

Valbonesi A, Ortenzi C, Luporini P (1992) The species problem in a ciliate with a high multiple mating system, *Euplotes crassus*. J Protozool 39: 45–54

Vallesi A, Giuli G, Bradshaw RA, Luporini P (1995) Autocrine mitogenic activity of pheromones produced by the protozoan ciliate *Euplotes raikovi*. Nature 376: 522–524

Von Heijne G (1987) Sequence analysis in molecular biology. Academic Press, New York

Watanabe T (1990) The role of ciliary surfaces in mating in *Paramecium*. In: Boodgood RA (ed) Ciliary and flagellar membranes. Plenum Press, New York, pp 149–171

Weischer A, Freiburg M, Heckmann K (1985) Isolation and partial characterization of gamone 1 of *Euplotes octocarinatus*. FEBS Lett 191: 176–180

Weiss MS, Anderson DH, Raffioni S, Bradshaw RA, Ortenzi C, Luporini P, Eisenberg D (1995) A cooperative model for ligand recognition and cell adhesion: evidence from the molecular packing in the 1.6 Å crystal structure of the pheromone E*r*-1 from the ciliate protozoan *Euplotes raikovi*. Proc Natl Acad Sci USA 92: 10172–10176

Wickner WT, Lodish HF (1985) Multiple mechanisms of protein interaction into and across membranes. Science 230: 400–407

Cell-Surface GPI Expression in Protozoa. The Connection with the PI System

P. Kovács[1]

1 Introduction

1.1 The GPI Anchor

Although the basic structure of biological membranes is provided by the lipid bilayer, most of the specific functions are carried out by proteins. Different membrane proteins are associated with the membrane in different ways. Results suggesting hydrophobic nonprotein anchors, e.g., glycosyl-phosphatidylinositol (GPI) were obtained in studies among others on membrane-bound enzymes, antigens, and cell-surface glycoproteins. It is somewhat surprising that a significant portion of proteins expressed at the cell surface (of both unicellular and higher eukaryotes) via GPI anchors, among others hydrolitic enzymes (e.g., aminopeptidase P, alkaline phosphatase, acetylcholinesterase, lipoprotein lipase); mammalian antigens (e.g., Thy 1, Q a, RT-G, CD 14); protozoal antigens [e.g., variant surface glycoprotein (VSG) (*Trypanosoma*), temperature-specific 156 G antigen (*Paramecium*), immobilization antigens (*Tetrahymena*), 195 kDa antigen (*Plasmodium*)]; cell adhesion molecules (e.g., N-CAM, heparan sulfate proteoglycan, guinea pig sperm PH20, contact side A (*Dictyostelium*), and another kind of proteins, such as 34-kDa placental growth factor, decay-accelerating factor (DAF), tegument protein (*Schistosoma*), 125-kDa glycoprotein (*Saccharomyces*), etc. (reviewed by Ferguson and Williams 1988).

The orientation of GPI-anchored proteins is extracellular; the treatment of the cells with phosphatidylinositol-specific phospholipase C (PI-PLC) or with glycosyl-phosphatidylinositol-specific phospholipase C (GPI-PLC) results in the release of these anchored proteins from the cell surface.

Because of their biological and medical significance, the GPI-anchored membrane proteins have become an area of intense investigation in recent years. For lack of normal synthesis of GPI or in the case of GPI-anchoring defects, some diseases will develop, such as paroxysmal nocturnal hemoglobinuria (caused by a somatic mutation of PIGA gene, that participates in GPI

[1]Department of Biology, Semmelweis University of Medicine, P.O. Box 370, 1445 Budapest, Hungary

anchor biosynthesis; Kawagoe et al. 1994). The African *Trypanosomes* are able to survive in the host's bloodstream by virtue of their dense cell-surface coat, which contains about 10 million copies of a GPI-anchored VSG (Ferguson et al. 1991). In several *Leishmania* species, GPIs are involved in cell invasion and in the survival in the parasitophorous vacuole of the host macrophage (Kelleher et al. 1995). In the pathophysiology of giardiasis, the GPI-anchored invariant antigen plays an important role. The *Toxoplasma gondii* expresses five major surface proteins attached to the membrane by GPI anchors (Tomavo et al. 1992). Some GPI-anchored proteins play a role in the evasion of the host defence systems: the release of the GPI-anchored glycoproteins may contribute to immune evasion of the worm *Schistosoma mansoni* (Sauma and Strand 1990); and also the GPI-degrading activity of *Plasmodium* is considered as a defense machinery (Braun-Breton et al. 1988).

The GPI-anchored structures also play an important role in cell signaling. Simultaneously with protein release (e.g., by GPI-PLC found in the liver membrane) runs the production of biologically active lipids such as 1,2-diacylglycerol and phosphatidic acid, which could cross the plasma membrane and affect intracellular metabolism. Recent experiments supported the involvement of a phospholipase-C-catalyzed hydrolysis of GPI-anchored structures in the mechanism of insulin action. The insulin-sensitive GPI appears to exhibit considerable similarity to the GPI protein anchor. However, the molecular size of the insulin-sensitive GPI is smaller, and some studies have suggested a cytoplasmic orientation of this complex (Saltiel and Cuatrecasas 1986).

1.2 The GPI Anchor Biosynthesis

The biosynthesis of GPI-anchored proteins involves the synthesis of a GPI anchor, and the synthesis of a protein with cleavable hydrophobic COOH-terminal sequence. The basic structure of GPI anchors is conserved between different eukaryotes, such as the mammalian and unicellular eukaryote cells. The comparison of the protozoan and nonprotozoan GPIs suggests that they all contain a common Manα1-4GlcNα1-6-*myo*Ins-1-PO_4-lipid structural motif (Fig. 1), but beyond this the glycans of the protein-linked and free GPIs diverge in structure.

The relative abundance of *Trypanosoma brucei* VSG has made this organism extremely useful for the study of GPI anchor biosynthesis (Masterson et al. 1989). The pathway based on the cell-free trypanosome membrane systems contains the following steps:

1. UDP-GlcNAc + PI $\rightarrow$ GlcNAc-PI.
2. GlcNAc-PI $\rightarrow$ de-N-acetylation $\rightarrow$ GlcN-PI.

Steps (1) and (2) occur on the cytoplasmic face of the endoplasmic reticulum.

3. Three α-mannose molecules are transferred onto GlcN-PI → Man_3GlcN-PI.
4. Man_3GlcN-PI + ethanolamine phosphate (derived from membrane-bound PE) → EtNPMan_3GlcN-PI (glycolipid A′).
5. Fatty acid remodeling reactions:
   a) sn-2-fatty acid is removed → glycolipid θ;
   b) glycolipid θ is myristoilated → glycolipid A″ (sn-1-stearoyl-2-myristoil glycerol);
   c) sn-1-stearoyl group is removed and replaced by myristic acid → glycolipid A.

The mature GPI precursor contains two myristic acid residues. Concomitant with the formation of glycolipid A is the acylation of the inositol ring, the formation of glycolipid C, the inositol-acylated version of the glycolipid A. This precursors are resistant to cleavage by PI-PLC.

The phosphoethanolamine moiety provides a bridge (amide bond) between the glycolipid and the protein, and is therefore an indispensable component of the GPI anchor.

The preformed GPI anchor is added to protein in the endoplasmic reticulum. The transfer of GPI precursor to the protein is very rapid. The first step in this transfer is the removal of a hydrophobic COOH-terminal GPI signal peptide of 17–23 amino acids in the case of VSG. The experimental data suggest a two-part consensus signal for GPI anchoring: a COOH-terminal hydrophobic sequence of minimum length (13 amino acids for DAF) and a cleavage/attachment domain requiring a small amino acid (e.g., Ala, Asp, Asn, Cys, or Ser) at the attachment site. The consensus signal for GPI anchor attachment depends on the nature of the proteins, such as hydrophobicity and conformation (reviewed by Lublin 1992).

1.3 Enzymes with Specificities for GPI Anchors

The bacterial PI-PLCs (e.g., *Bacillus cereus, B. thuringiensis, Clostridium novyi*) show specificity for phosphoinositides but no other phospholipid molecules. These PI-PLCs will cleave most protein GPI anchors and release the proteins in a soluble form from the cell membrane. However, several GPI-anchored proteins are largely, sometimes completely resistant to PI-PLC. For example, the 5′-nucleotidase release requires 100 times more PI-PLC than alkaline phosphatase.

The GPI-PLC enzymes show a marked specificity for GPI anchors. In *Trypanosomes* this enzyme behaves as a nonglycosylated amphiphilic membrane protein. The specificity requirement of GPI-PLC involves the carbohydrate but not the protein or the fatty acids of GPI. A similar enzyme has been purified from rat liver membranes. These enzymes show no Ca^{2+} dependence,

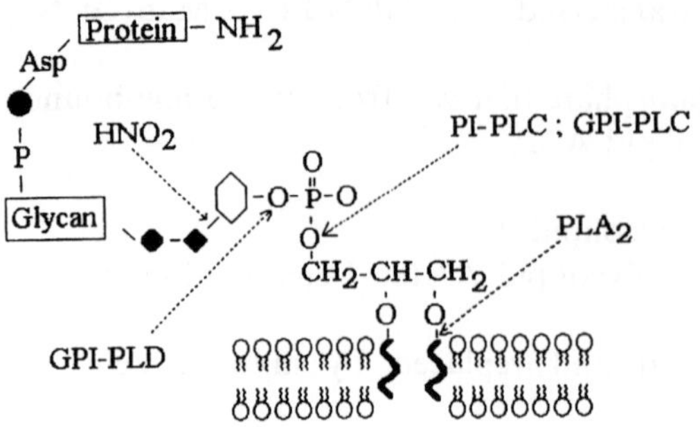

Fig. 1. Structural features of GPI anchors. *Arrows* indicate cleavage sites. ⬡ inositol; ◆ N-acetyl glucosamine; ● mannose; ● ethanolamine

and no activity toward the PIP and PIP_2, thus they are clearly different from the PI-PLCs involved in IP_3 generation.

The mammalian serum and the α and β cells of Langerhans islets contain high amounts of GPI-PLD, this enzyme is capable of degrading the GPI anchor cleaving the bonds between the inositol and phosphate, but have no activity towards PIP and PIP_2.

The PLA_2 enzymes (both cytosolic and secretory) can generate arachidonic acid and lysophospholipids, as LPA. This enzyme is capable of liberating the GPI-anchored proteins via cleavage of the ester bonds between the glycerol and fatty acids (Fig. 1; reviewed by Ferguson and Williams 1988).

1.4 The Inositol Phospholipids and Signal Transduction

A wide variety of extracellular signals effect responses in their target cells via activation of a different kind of second messenger systems. One such system is the inositol phospholipid signaling system.

The binding of a signaling molecule to a G-protein-linked receptor in the plasma membrane leads to the activation of phosphoinositide-specific phospholipase C (PI-PLC), which cleaves PIP_2 to generate the second messengers diacylglycerol (DAG) and inositoltrisphosphate (IP_3). These second messengers result in cellular stimulation via activation of protein kinase C [DAG] and elevation of intracellular Ca^{2+} (IP_3) (Berridge 1984).

More than 25 different cell-surface receptors have been shown to utilize this signaling pathway, among others hormones, such as vasopressin, acetycholin, and serotonin; moreover, some different cells and tissues respond to different kinds of external stimuli by PIP_2 breakdown, such as photons (photoreceptors), spermatozoa (sea urchin egg), glucose (pancreatic islets), and antigens (B cells).

The synthesis of inositol phospholipids means two cycles: the inositol phosphate cycle and the lipid cycle (Fig. 2). These cycles are connected with the production of other biologically active compounds, such as the arachidonic

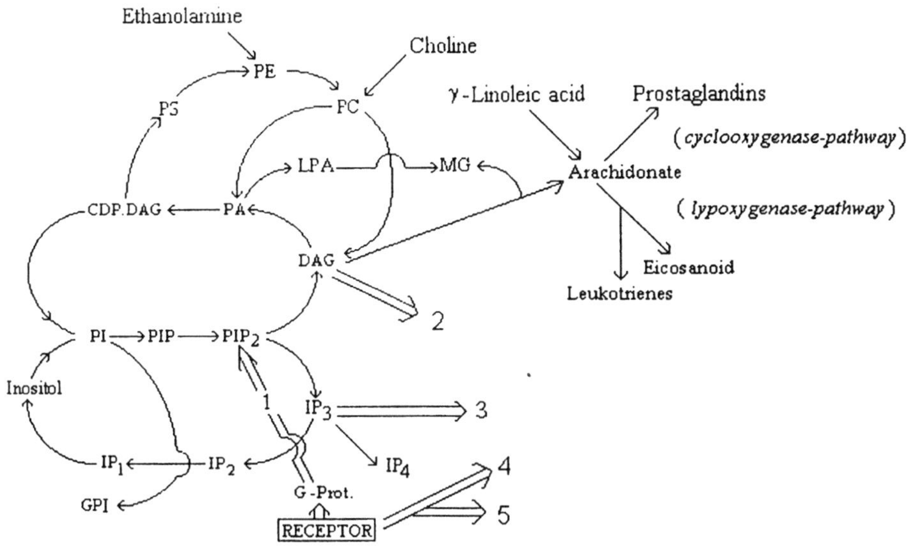

Fig. 2. Inositol phospholipid metabolism and its connections with the receptors and signal transduction. (After Berridge and Irvine 1984; Sekar and Hokin 1986). *1* Receptor-mediated breakdown of PIP_2 (via G-proteins); *2* effect of DAG on protein kinase C; *3* effect of IP_3 on the Ca^{2+} and the Ca-calmodulin system; *4,5* effect of G-proteins on the cAMP and cGMP level. Abbreviations: *PA* phosphatidic acid; *LPA* lysophosphatidic acid; *PC* phosphatidyl choline; *PE* phosphatidyl ethanolamine; *PS* phosphatidyl serine; *DAG* diacylglycerol; *MG* monoacylglycerol; *CDP-DAG* cytidine-diphosphate diacylglycerol; *PI* phosphatidylinositol; *PIP* phosphatydilinositol 4-monophosphate; PIP_2 phosphatidylinositol 4,5-bisphosphate; IP_3 inositol 1,3,4-trisphosphate; IP_4 inositol 2,3,4,6-tetrakisphosphate; IP_2 inositol 1,4-bisphosphate; IP_1 inositol 1-monophosphate

acid metabolites. On the other hand, a link between the activity of the PI synthesis and the synthesis of the GPI anchor can be supposed (Berridge and Irvine 1984).

The phosphoinositides are minor constituents of both unicellular and higher eukaryote cells and are located in the plasma membrane. The role of phosphoinositides in signal transduction in higher eukaryotes is well documented, but in unicellular organisms is not well understood.

2 The Cell-Surface Expression of GPI-Anchored Proteins in the Protozoa

2.1 GPI-Anchored Proteins in the Parasitic Protozoa

While mammalian cells typically express in the order of 100 thousand copies of GPI anchor per cell, the parasitic protozoa, particularly the kinetoplastids, express up to 10–20 million copies of GPI anchor and/or GPI-related glycolipids per cell (Ferguson et al. 1994). In several cases, the GPI-anchored molecules are known to be essential for parasite survival and infectivity.

The highly elevated levels and the specialized nature of GPI metabolism in these parasites suggest that the GPI biosynthetic pathways might be good targets for the development of chemotherapeutic agents.

In the protozoa, the most highly expressed GPI-anchored proteins appear to be important molecules in the protection of the cells and the nutrient uptake, and show – besides the very high surface expression – very low turnover rates (Ferguson et al. 1991).

Trypanosoma brucei. The GPI anchor of trypanosome VSG is unusual in that its fatty acids are exclusively myristate. The myristate is added to GPI in a fatty acid remodeling reaction forming glycolipid A. The *Trypanosoma* have a second myristate exchange pathway which results in a lyso-GPI, named theta (Θ), which has myristate as its sole fatty acid. The myristate exchange and fatty acid remodeling appear to occur in different subcellular compartments and the two reactions have different sensitivities to inhibitors (Buxbaum et al. 1994). The myristate analogs are useful for studying the mechanism of GPI myristoilation, and because of some analogs, as the 11-oxatetradecanoic acid are toxic for *Trypanosoma*, are candidates for antitrypanosomal chemotherapy (Doering et al. 1994).

The compound 2-deoxy-2-amino-D-mannose (mannosamine) has been shown to be an inhibitor of N-glycosylation (Pan and Elbein 1985) and GPI anchor precursor formation in both mammalian cells and *Trypanosoma brucei* procyclic form via incorporating into GPI anchor intermediates (Lisanti et al. 1991). Another possibility for the inhibition of GPI anchor biosynthesis is the serine esterase inhibitor phenylmethanesulphonyl fluoride (PMSF), which inhibits phosphoethanolamine incorporation into the GPI anchor precursor, resulting in the accumulation of a Man_3GlcNH_2-PI intermediate (Masterson and Ferguson 1991). The *Trypanosoma brucei*-purified GPI derived from VSG induces rapid onset of tyrosine phosphorylation of multiple intracellular substances by activation of protein tyrosine-kinase. This activation is followed by cytokine (IL-1) expression in macrophages. Thus, GPI induces cytokine excess, causing certain pathological states associated with trypanosomiasis (Tachado and Schofield 1994).

Trypanosoma brucei VSG are cleaved by the GPI-PLC of the trypanosome membrane. The loss of VSG from bloodstream forms occurs during differentiation to procyclic forms. However, evidence that GPI-PLC is involved in VSG metabolism in the living trypanosome is circumstantial; GPI-PLC is found on the cytoplasmic face of the vesicles, so through vesicle fusion it could not gain access to the VSG. The loss of VSG occurs through the action of protease (Webb et al. 1994).

An inositol glycan fragment from the trypanosome VSG anchor shows insulin-mimetic antilipolytic activity, e.g., dose-dependently inhibits isoproterenol-stimulated lipolysis (Misek and Saltiel 1992).

Trypanosoma cruzi. *Trypanosoma cruzi* invades the host cells by a mechanism distinct from phagocytosis. During cell invasion, *Trypanosoma cruzi* recruits host lysosomes, which gradually fuse with the plasma membrane at the site of parasite entry. Thus, the membrane of the parasitophorous vacuole is very similar to the membrane of lysosomes. The major glycoproteins (Lgps) from mammalian lysosomes are desialylated by a GPI-anchored *trans*-sialidase present on the surface of trypomastigote forms. The desialyzation of Lgps facilitates membrane lysis by a parasite-secreted molecule (Tc-TOX), which has membrane pore-forming activity (Andrews 1994). The members of the GP85 sialidase family of *Trypanosoma cruzi* show several amino acid motifs, among others the putative GPI anchor sequence. Sequence analysis revealed that the GP82-stage-specific surface antigens have similar amino acid motifs, thus also this antigen belongs to the sialidase family. This antigen is encoded by a multigen family whose members are distributed in several chromosomes, and the mRNAs for the GP82 are developmentally regulated (Araya et al. 1994).

The surface of *Trypanosoma cruzi* contains GPI-anchored mucin-like molecules, which function as sialic acid acceptors. The *trans*-sialidase catalyzes the transfer of sialic acid from the host to the *Trypanosoma cruzi* GPI-anchored surface acceptors. Interestingly, the mucins isolated from the insect stages differ from those isolated from the mammalian stages in reaction to monoclonal antibodies (Acosta et al. 1994).

Trypanosoma cruzi expresses GPI-anchored membrane antigen with cysteine proteinase activity (GP 50/55), and this antigen differs from lysosomal cystein proteinase (GP 57/51), showing more homology to papain (prototype of the cysteine proteinases) than lysosomal GP 57/51 (Fresno et al. 1994).

Giardia lamblia. *Giardia lamblia* is a major cause of waterborne enteric disease throughout tropical zones. This primitive protozoan expressing GPI-anchored surface proteins (GP 49) was found to be present in different isolates of *Giardia*, and thus can be considered as an invariant antigen. The intestinal fatty acids may become incorporated into *Giardia* via GP 49, moreover this GPI-anchored protein regulates the fluid secretion of intestinal epithelial cells by the alteration of the electrolyte fluxes (Das et al. 1994).

Leishmania mexicana; *L. major*. The amastigotes (mammalian stage) of *Leishmania* proliferate in phagolysosomes of macrophages. They express glycoinositol phospholipids (GIPLs), which are necessary for parasite survival by providing a shield at the cell surface against lysosomal hydrolases and by serving as receptors for interaction with host cells. The structural analyses show that the GIPL glycans (e.g., Manα1-3Manα1-4GlcN in the iM_2) are linked to alkylacyl-phosphatidylinositol. Thus, the GIPLs of *Leishmania* are structurally related to the GPI anchors of both the surface proteins and the lipophosphoglycans (LPGs; Schneider et al. 1994). The lipophosphoglycans (LPGs) are expressed in high copy number of the promastigote (insect-dwelling) stage, but are in very low levels in the amastigote stage. The LPG from *Leishmania major* contains β-D-Arap-terminating side chains that are involved in regulating the attachment of the parasite to the midgut epithelium of its insect vector. *L. major* invades mononuclear cells by an interaction between the surface LPGs and macrophage receptors. This interaction is mediated by phosphoglycan repeats containing oligomers of β(1–3)Gal residues. Although amastigotes also use LPGs to bind to the macrophages, this interaction is independent of the β(1–3)Gal residues (Kelleher et al. 1995). These molecules are downregulated during transformation from the promastigote to the amastigote stage, suggesting that they have evolved for survival in the insect vector (Ferguson et al. 1991).

The LPGs and GIPLs are not linked to protein, and there are no known counterparts in other eukaryotes.

2.2 *GPI-Anchored Proteins in Free-Living Protozoa*

Recently, several papers have described the occurrence of GPI-anchored proteins in various species of the ciliated protozoans *Tetrahymena* and *Paramecium*. However, elucidation of the functional significance of the GPI anchors has lagged significantly in comparison with the detailed structural studies performed.

The inositol glycolipids have structural features in common with the other GPIs, GIPLs, and LPGs that have been characterized, e.g., in *Leishmania* and *Trypanosoma*.

Tetrahymena mimbres; T. thermophyla; T. vorax. The structure of the glycan head group of the *Tetrahymena mimbres* glycolipids is unique compared with the GPI structures that have been determined in other species:

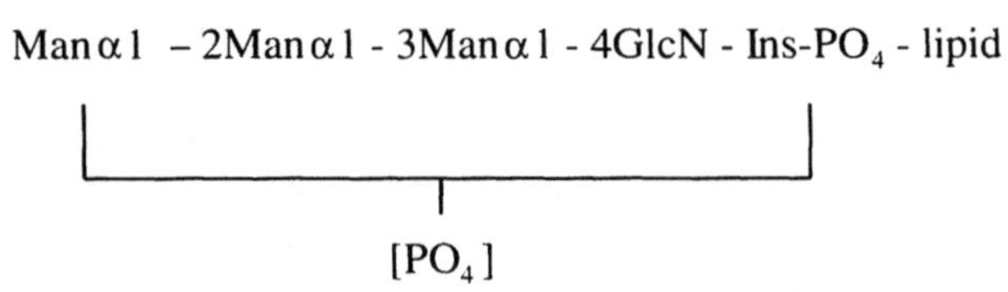

(Weinhart et al. 1991). The inositol lipid appears to be sn-1-alkyl, 2-lysophosphatidylinositol, perhaps with some sn-1-acyl, 2-lysophosphatidylinositol also being present. An additional, possible modification is the esterification of a fatty acyl chain to one of the available hydroxyl groups of the inositol ring, but it probably does not account for a major proportion of the GPI glycan, since much of the material is susceptible to PI-PLC attack, and the GPI glycans esterified e.g., with palmitic acid are PI-PLC-resistant (Ryals et al. 1991).

The examination of the dynamics of GPI-anchored protein release is problematic. Because *Tetrahymena* secretes a variety of potent hydrolytic enzymes into its surrounding medium (Müller 1972), precautions are necessary in order to prevent rapid degradation of the proteins released from the cell surface.

Experiments show that in *Tetrahymena mimbres*, the apparent molecular weights of the GPI-anchored proteins have temperature-induced differences. These slightly different proteins may be expressed at different temperatures, or may be subject to posttranslational modifications affected by temperature. The temperature-specific GPI-anchored proteins are in all probability immobilization antigens (i-antigens). This likelihood is strengthened by the examination of the *Paramecium* temperature-specific i-antigens, which are linked to the cell surface by GPI glycan (Azzouz and Capdeville 1992), according to experiments carried out by Ko and Thompson (1992) on *Tetrahymena thermophila*. For example, the H3 antigen of *T. thermophila* is not produced by cells growing at 40 °C, but is rapidly induced after shifting cells to 28 °C. The newly induced H3 antigens appear on the surface of these cells within 30 min, first on the nonciliated regions, and then gradually along the surface of the cilia. Evidence reported by Bolivar and Guiard-Maffia (1989) indicates that the antigens may be transported to the cell surface as mucocyst secretions, and some of the material may subsequently be internalized during the formation of food vacuoles at the cytopharynx. This likelihood of continuous rapid secretion and shedding of the GPI-anchored proteins may explain the relatively low abundance of the i-antigens associated with the outer surface at any given time. The susceptibility of the i-antigen-GPI to hydrolysis by endogenous PLCs raises the possibility that such a cleavage is a normal step in i-antigen release (Ko and Thompson 1992).

Tetrahymena vorax is capable of differentiating into an alternative phenotype in response to an external chemical signal. The differentiation involves significant reorganization of the cell's internal membranes and restructuring of the cytoskeletal architecture. The differentiation can be induced by water-soluble exudate (e.g., stomatin) obtained from prey organisms (Buhse 1967). Under standard culture conditions, *T. vorax* contains three GPI-anchored polypeptides. Some data indicate that the ability of *T. vorax* populations to differentiate depends on continuous GPI-anchored protein assembly, and that a new GPI-anchored protein appears during the differentiation event; the GPI-anchored protein assembly is a probable requirement for differentiation. In the presence of exogenous PI-PLC, the differentiation is accelerated. A possible interpretation of this phenomenon would be that the GPI-anchored proteins themselves are a target of the differentiation process, and the presence of exogenous PI-PLC accelerates the normal sequence of events (Yang and Ryals 1994). The data suggest an exciting possibility that a transmembrane- signaling cascade (the binding of stomatin to the GPI-anchored receptor proteins followed by internalization) is involved in the differentiation. The parasomal sacs of *Tetrahymena* might function as primitive analogs of caveolae, and the signaling starts with the phenomenon of caveolae-mediated potocytosis.

Another possible signaling role of *Tetrahymena piriformis* GPI is indicated by the effects of insulin treatments. *T. piriformis* responds to insulin by alteration of its sugar metabolism, phagocytotic activity, insulin-binding capacity, behavior, etc. (reviewed by Kovács 1986; Csaba 1994). GPI is considered as a putative second messenger of insulin (Saltiel and Cuatrecasas 1986). Insulin treatment (imprinting; Csaba, this Vol.), and especially the second treatment of imprinted cells by 10^{-6} M insulin, which significantly elevated the amount of ^{3}H-inositol incorporated into GPIs, indicate that GPIs are working in *Tetrahymena* and may be connected with insulin action (Kovács and Csaba 1994a). This conclusion is based on the fact that the GPI of *T. pyriformis* could be labeled in vivo with a variety of radioactive compounds known to be present in GPI, including inositol, glucosamine, mannose, and ^{32}P (Kovács and Csaba 1995a); but the existence and structure of putative *Tetrahymena* glycolipid cleaved from the GPI remain to be determined.

3 Inositol Phospholipids in *Tetrahymena pyriformis*. The Possible Link Between the PI System and Synthesis of GPI

Phosphoinositides are minor constituents of eukaryotic cells, and are located in the plasma membrane. Phosphatidylinositol (PI) is the most abundant representative, whereas phosphatidylinositol 4-phosphate (PIP) and phosphatidylinositol 4,5-bisphosphate (PIP_2) often occur in trace amounts only (Michell 1975).

Tetrahymena species appear to be fully capable of synthesizing all their phospholipids (Smith 1993). PI has been found as a component of the glycosylphosphatidyl inositol anchor for cell surface proteins (Ryals et al. 1991; Weinhart et al. 1991; Ko and Thompson 1992; Yang and Ryals 1994), but free PI constitutes a small portion of the *Tetrahymena* phospholipids (Kovács and Csaba 1990; Smith 1993).

Besides the function of PI as a component of GPI, the role of phosphoinositides in the unicellular organisms is not well understood. In *Trypanosoma cruzi*, various forms of phosphoinositides were found, and the presence of PIP and PIP_2 would indicate the presence of kinases responsible for PI phosphorilation (Racagni et al. 1992), but the biological significance of metabolism of phosphoinositides was not discussed.

In the wall-less mutant strain of *Neurospora crassa*, long-term (20-h) insulin treatment increased the amount of PIP_2 in the insulin-pretreated (imprinted) group, although short (30 s) insulin treatment had no affect on it (László and Csaba 1992). Several studies showed that pretreatment (imprinting) with a hormone changes the binding capacity and thus the degree of hormone effect in a subsequent treatment (Csaba 1985). Such an effect was shown in *Tetrahymena*, when, even on the level of phospholipids, the quantity of PIP_2 decreased in pretreated cells, and without pretreatment there were no changes (Kovács and Csaba 1990). In *Neurospora*, although long pretreatment increased the amount of PIP_2, these changes play no role in insulin signal transduction (László and Csaba 1992).

In *Tetrahymena pyriformis*, supplementing the culture medium with choline chloride and ethanolamine (10^{-3} M) increased the PI, PIP, and PIP_2 levels.

These results indicate that those molecules in the environment which could be considered to be likely precursors of a second messenger system are capable of influencing the intracellular level of the second messengers (Kovács and Csaba 1995b).

AlF_4 and BeF_3 complexes, which are formed in the simultaneous presence of $AlCl_3$ or $BeCl_2$ and NaF, are able to activate the heterotrimer G proteins. The complexes have the capacity to express similar effects on phospholipase C as the γ-phosphate of GTP (Antonny and Chabre 1992). In this way, treatment with AlF_4 and BeF_3 complexes could alter the quantities and the ratios of individual members of the PI system. Experiments with these G protein-activating fluorides suggest the presence of the G proteins in *Tetrahymena pyriformis*, and indicate the function of a G protein-regulated PI system (Kovács and Csaba 1994b). Treatment with 10 mM NaF alone or in combination with $BeCl_2$ or $AlCl_3$ made possible a rapid ^{32}P incorporation and appearance in members of the PI system, and these treatments increased the ratio of the PI system compared to the total phospholipid content.

Besides the effect on PI metabolism, in *Tetrahymena* the fluorides influenced dose-dependently the insulin-binding capacity and the imprintability with insulin (Kovács and Csaba 1995c). Similarly, treatment of *Tetrahymena* cells by protein kinase C-activating phorbol esters increased ^{32}P incorporation into the inositol phospholipids (Kovács and Csaba 1995d), and was followed by elevated insulin-binding capacity (Kovács and Csaba 1995c). There appears to be a connection in *Tetrahymena* between the intensity of PI metabolism and the synthesis of GPI. In the presence of phenothiazines (chlorpromazine, propericiazine, trifluoperazine), the incorporation of ^{32}P into the inositol phospholipids decreased dramatically, and also GPI synthesis diminished significantly. Treatment with local anesthetics (tetracaine, procaine) caused a slightly elevated ^{32}P incorporation into the inositol phospholipids, and resulted in an elevated GPI level compared to the control cells (Kovács and Csaba 1995e). On the other hand, the level of GPI in *Tetrahymena* is probably connected with the insulin-binding capacity: the treatment of *Tetrahymena* with local anesthetics significantly increased the amount of membrane-bound insulin (Nozawa et al. 1985).

We assume that, similarly to the environmentally controlled expression of different GPI-anchored antigenic variations, hormonal (insulin) treatments alter the level and synthetic activity of GPI with the help of metabolic activity of the PI system. The altered GPI level may account partly for the altered surface-bound glycoproteins, and partly for the putative insulin-sensitive inositol glycans, which, as a second messenger, could mediate the action of insulin.

References

Acosta A, Schenkmann RP, Schenkmann S (1994) Sialic acid acceptors of different stages of *Trypanosoma cruzi* are mucin-like glycoproteins linked to the parasite membrane by GPI anchors. Braz J Med Biol Res 27: 439–442

Andrews NW (1994) From lysosomes into the cytosol: the intracellular pathway of *Trypanosoma cruzi*. Braz J Med Biol Res 27: 471–475

Antonny B, Chabre M (1992) Characterization of the aluminium and beryllium fluoride species which activate transducin. J Biol Chem 267: 6110–6118

Araya JE, Cano MI, Yoshida N, da-Silveira JF (1994) Cloning and characterization of a gene for the stage-specific 82 kDa surface antigen of metacyclic trypomastigotes of *Trypanosoma cruzi*. Mol Biol Parasitol 65: 161–169

Azzouz N, Capdeville Y (1992) Structural comparisons between the soluble and the GPI-anchored forms of the *Paramecium* temperature-specific 156 G surface antigen. Biol Cell 75: 217–223

Berridge MJ (1984) Inositol trisphosphate and diacylglycerol as second messengers. Biochem J 220: 345–360

Berridge MJ, Irvine RF (1984) Inositol trisphosphate, a novel second messenger in cellular signal transduction. Nature 312: 315–321

Bolivar I, Guiard-Maffia J (1989) Cellular localization of the SerH surface antigen in *Tetrahymena thermophila*. J Cell Sci 94: 343–354

Braun-Breton C, Rosenberry TL, da Silva LP (1988) Induction of the proteolytic activity of a membrane protein in *Plasmodium falciparum* by phosphatidyl inositol-specific phospholipase C. Nature 332: 457–459

Buhse HE Jr (1967) Microstome–macrostome transformation in *Tetrahymena vorax* strain V2 type S induced by a transforming principle, stomatin. J Protozool 14: 608–613

Buxbaum LU, Raper J, Opperdoes FR, Englund PT (1994) Myristate exchange. A second glycosyl phosphatidylinositol myristoilation reaction in African trypanosomes. J Biol Chem 269: 30212–30220

Csaba G (1985) The unicellular *Tetrahymena* as a model cell for receptor research. Int Rev Cytol 95: 327–377

Csaba G (1994) Phylogeny and ontogeny of chemical signaling: origin and development of hormone receptors. Int Rev Cytol 155: 1–48

Das S, Traynor-Kaplan A, Kachintorn U, Aley SB, Gillin FD (1994) GP 49, an invariant GPI-anchored antigen of *Giardia lamblia*. Braz J Med Biol Res 27: 463–469

Doering TL, Lu T, Werboretz KA, Gokel GW, Hart GW, Gordon JI, Englund PT (1994) Toxicity of myristic acid analogs toward African trypanosomes. Proc Natl Acad Sci USA 91: 9735–9739

Ferguson MA, Williams AF (1988) Cell surface anchoring of proteins via glycosyl-phosphatidyl-inositol structures. Annu Rev Biochem 57: 285–320

Ferguson MA, Masterson WJ, Homans SW, McConville MJ (1991) Evolutionary aspects of GPI metabolism in kinetoplastid parasites. Cell Biol Int Rep 15: 991–1005

Ferguson MA, Brimacombe JS, Cottaz S, Field RA, Guther LS, Homans SW, McConville MJ, Mehlert A, Milne KG, Ralton JE (1994) Glycosyl-phosphatidylinositol molecules of the parasite and the host. Parasitology (Suppl) 108: 45–54

Fresno M, Hernandez-Murain C, de-Diego J, Rivas L, Scharfstein J, Bonay P (1994) *Trypanosoma cruzi*: identification of a membrane cystein proteinase linked through a GPI anchor. Braz J Med Biol Res 27: 431–437

Kawagoe K, Takeda J, Endo Y, Kinoshita T (1994) Molecular cloning of murine pig-a, a gene for GPI-anchor biosynthesis, and demonstration of interspecies conservation of its structure, function and genetic locus. Genomics 23: 566–574

Kellcher M, Moody SF, Mirabile P, Osborn AH, Bacic A, Handman E (1995) Lipophosphoglycan blocks attachment of *Leismania major* amastigotes to macrophages. Infect Immun 63: 43–50

Ko YG, Thompson GA Jr (1992) Immobilization antigens from *Tetrahymena thermophila* are glycosyl-phosphatidylinositol-linked proteins. J Protozool 39: 719–723

Kovács P (1986) The mechanism of receptor development as implied from hormonal imprinting studies on unicellulars. Experientia 42: 770–775
Kovács P, Csaba G (1990) Involvement of the phosphoinositol (PI) system in the mechanism of hormonal imprinting. Biochem Biophys Res Commun 170: 119–126
Kovács P, Csaba G (1994a) Effect of insulin on the incorporation of ^{3}H-inositol into the inositol phospholipids (PI, PIP, PIP_2) and glycosyl-phosphatidylinositols (GPIs) of *Tetrahymena pyriformis*. Biosci Rep 14: 215–219
Kovács P, Csaba G (1994b) Effect of G-protein activating fluorides (NaF, AlF_4 and BeF_3) on the phospholipid turnover and the PI system of *Tetrahymena*. Acta Protozool 33: 169–175
Kovács P, Csaba G (1995a) The effects of aminosugars (glucosamine, mannosamine) and 2-deoxy-fluoroglucose on the phosphatidyl inositol (PI) and glycosyl phosphatidylinositol (GPI) systems of *Tetrahymena*. Microbios (in press)
Kovács P, Csaba G (1995b) Effects of choline and ethanolamine on the synthesis and breakdown of the inositol phospholipid (PI) system in *Tetrahymena*. Cell Biochem Funct 13: 61–67
Kovács P, Csaba G (1995c) Effects of G-protein activator fluorides, protein kinase C activator phorbol ester and protein kinase inhibitor on insulin binding and hormonal imprinting of *Tetrahymena*. Microbios 81: 231–239
Kovács P, Csaba G (1995d) Effect of phorbol 12-myristate 13-acetate (PMA) on the phosphoinositol (PI) system in *Tetrahymena*. Study of ^{32}P incorporation and breakdown of phospholipids. Cell Biochem Funct 13: 85–89
Kovács P, Csaba G (1995e) The effects of local anesthetics and phenothiazines on ^{32}P incorporation into the phosphoinositides and GPI of *Tetrahymena pyriformis*. Acta Protozool (in press)
László V, Csaba G (1992) Phospholipid content of untreated and insulin treated cells of *Neurospora crassa* (wall-less mutant strain). Acta Microbiol Hung 39: 229–233
Lisanti MP, Field MC, Caras IWJ, Menon AK, Rodriquez-Boulan E (1991) Mannosamine, a novel inhibitor of glycosylphosphatidylinositol incorporation into proteins. EMBO J 10: 1969–1977
Lublin DM (1992) Glycosyl-phosphatidylinositol anchoring of membrane proteins. Curr Topics Microbiol Immunol 178: 141–162
Masterson WJ, Ferguson MA (1991) Phenylmethanesulfonyl fluoride inhibits GPI anchor biosynthesis in the African trypanosome. EMBO J 10: 2041–2045
Masterson WJ, Doering TL, Hart GW, Englund PT (1989) A novel pathway for glycan assembly: biosynthesis of the glycosyl-phosphatidylinositol anchor the trypanosome variant surface glycoprotein. Cell 56: 793–800
Michell RH (1975) Inositol phospholipids and cell surface receptor function. Biochem Biophys Acta 415: 81–147
Misek DE, Saltiel AR (1992) An inositol phosphate glygan derived from a *Trypanosoma brucei* glycosyl-phosphatidylinositol mimics some of the metabolic actions of insulin. J Biol Chem 267: 16266–16273
Müller M (1972) Secretion of acid hydrolases and its intracellular source in *Tetrahymena pyriformis*. J Cell Biol 52: 478–487
Nozawa Y, Kovács P, Csaba G (1985) The effects of membrane perturbants, local anesthetics and phenothiazines on hormonal imprinting in *Tetrahymena pyriformis*. Cell Mol Biol 31: 223–227
Pan YT, Elbein AD (1985) The effect of mannosamine on the formation of lipid-linked oligosaccharides and glycoproteins in canine kidney cells. Arch Biochem Biophys 242: 447–456
Racagni G, Garcia de Lema M, Domenech CE, Machado de Domenech EE (1992) Phospholipids in *Trypanosoma cruzi*: phosphoinositide composition and turnover. Lipids 27: 275–278
Ryals PE, Pak Y, Thompson GA Jr (1991) Phosphatidylinositol-linked glycans and phosphatidylinositol-anchored proteins of *Tetrahymena mimbres*. J Biochem 266: 15048–15053
Saltiel AR, Cuatrecasas P (1986) Insulin stimulates the generation from hepatic plasma membranes of modulators derived from an inositol glycolipid. Proc Natl Acad Sci USA 83: 5793–5797
Sauma SY, Strand M (1990) Identification and characterization of glycosylphosphatidylinositol-linked *Schistosoma mansoni* adult worm immunogens. Mol Biochem Parasitol 38: 199–209
Schneider P, Schnur LF, Jaffe CL, Ferguson MA, McConville MJ (1994) Glycoinositol-phos-

pholipid profiles of four serotypically distinct Old World *Leishmania* strains. Biochem J 304: 603–609

Sekar MC, Hokin LE (1986) The role of phosphoinositides in signal transduction. J Membr Biol 89: 193–210

Smith JD (1993) Phospholipid biosynthesis in protozoa. Prog Lipid Res 32: 47–60

Tachado SD, Schofield L (1994) Glycosylphosphatidylinositol toxin of *Trypanosoma brucei* regulates IL-1 α and TNF-α expression in macrophages by protein tyrosine kinase mediated signal transduction. Biochem Biophys Res Commun 205: 984–991

Tomavo S, Dubrenetz JF, Schwarz RT (1992) Biosynthesis of glycolipid precursors for glycosylphosphatidylinositol membrane anchors in a *Toxoplasma gondii* cell free system. J Biol Chem 267: 21446–21458

Webb H, Carnoll N, Carrington M (1994) The role of GPI-PLC in *Trypanosoma brucei*. Braz J Med Biol Res 27: 349–356

Weinhart U, Thomas JR, Pak Y, Thompson GA Jr, Ferguson MA (1991) Structural characterization of a novel glycosyl-phosphatidylinositol from the protozoan *Tetrahymena mimbres*. Biochem J 279: 605–608

Yang X, Ryals PE (1994) Cytodifferentiation in *Tetrahymena vorax* is linked to glycosyl-phosphatidylinositol-anchored protein assembly. Biochem J 298: 697–703

Cell Adhesion Proteins in the Nonvertebrate Eukaryotes

P.N. Lipke[1]

1 Introduction

1.1 History and Philosophy

It is an interesting time to consider the means and roles of cell adhesion in the eukaryotes. There was initial wonder at the results of Wilson's reports of species-specific reaggregation of sponge cells (Wilson 1907), and Holtfreter's demonstrations of cell segregation in amphibians (Holtfreter 1943). Later, there were theories about cell adhesion molecules driving histogenesis and differentiation, and speculation about how specificity could be generated (Tyler 1947; Weiss 1947; Roseman 1970). The rise of molecular methods in the 1950s and 1960s revealed that the field was extremely confusing. Various research groups could not agree on results from similar experiments, let alone on their interpretation. This confusion, in fact, stemmed from an unappreciated complexity, with the same cells adhering by different mechanisms, depending on details of cell preparation, timing, and criteria for assay endpoints. It was a great revelation that chick neural cells could adhere to each other in two different reproducible ways, depending on the concentration of Ca^{2+} present during tissue dissociation (Takeichi et al. 1981). In the past 10 years, we have developed frameworks and rules for understanding cell adhesion at a molecular level, aided by molecular genetics and the ability to compare gene sequences and functions. The emergence of identifiable gene families provides a frame of reference to describe the behavior of cell adhesion molecules. At the same time, molecular evolution methods have clarified the relationships among the eukaryotic kingdoms and phyla, and so we can now begin to trace the evolutionary history of the genes, as well as the organisms.

Moscona's discovery that chick and mouse liver cells adhered together while chick liver and brain cells segregated was a surprising result (Moscona 1957). Moscona recognized that histotypy of adhesion meant that some molecular characteristics and binding sites were the same in mammals and birds, and that, therefore, the specificity problem had been solved early in vertebrate evolution.

---

[1]Department of Biological Sciences and the Institute for Biomolecular Structure and Function, Hunter College of the City University of New York, New York, NY 10021, USA

This discovery foretold the realization that cell adhesion molecules are members of gene families and superfamilies present before the invertebrate/vertebrate split, and in some cases, before the divergence of the multicellular eukaryote kingdoms. These gene families encode cell adhesion molecules with conserved cellular roles as well as molecular structures, and we can now begin to make some sense of the plethora of information on cell adhesion. In addition, we now find recurring motifs of structure and function within these gene families: structural domains, secretion signals, cell anchorage signals, signal transduction domains, and domains of recognizable similarity but unknown function.

1.2 Evolution

This essay is organized around an evolutionary tree derived from rRNA sequences (Fig. 1; Patterson and Sogin 1992; Hinkle et al. 1994). Analyses of the divergence of these sequences show that the eukaryote kingdoms are in some sense nonequivalent: the protista include organisms of both ancient and modern origin, while the multicellular kingdoms: plants, fungi, and animals,

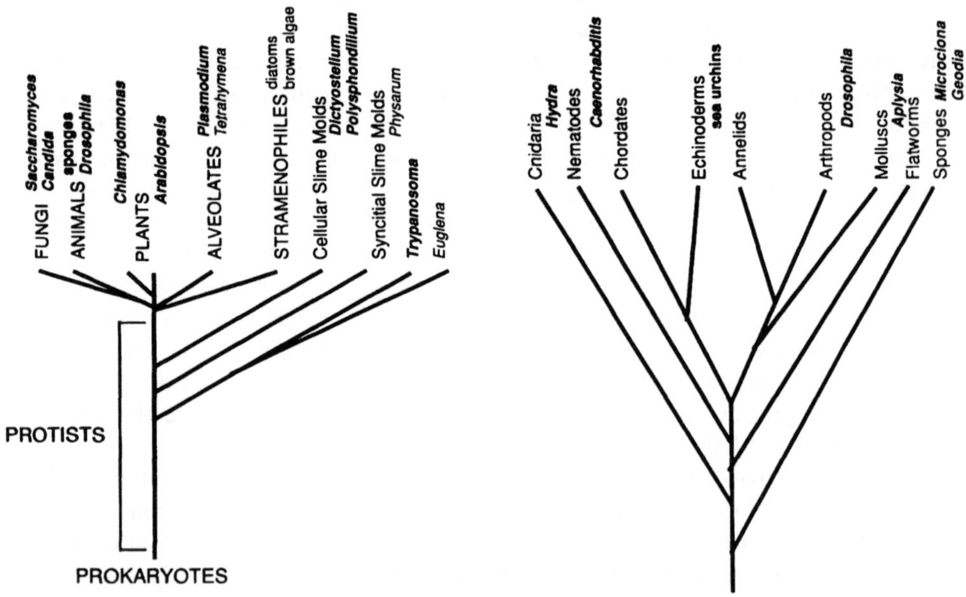

Fig. 1. Evolutionary relationships among organisms used in cell adhesion studies. Organisms discussed in detail are shown in *boldface*. The branching order of various taxons is shown, as determined from electron microscopy and rRNA sequences (Patterson and Sogin 1992). The oldest divergences are shown at the *bottom*. *Left* Relationships among the major kingdoms; *right* relationships of major animal phyla

share a common and relatively recent origin in the "Precambrian explosion," about 10^9 years ago. There are five major groups that derive from this explosion: the animals, green plants, fungi, alveolates (such as *Plasmodium* and *Tetrahymena*), and stramenophiles (including, among others, oomycetes, diatoms, and brown algae). Accordingly, many animal cell adhesion molecules have sequence similarities that imply evolutionary origins prior to the Precambrian divergence. For instance, some members of the immunoglobulin (Ig) and integrin families of cell adhesion glycoproteins are present in fungi as well as in animals, and fasciclin I-like sequences are found in humans, flies, and algae. In contrast, the protists represent a collection of organisms that are primarily unicellular, but contain some taxons that are extremely ancient and some that are modern. As a consequence of the diversity of the kingdom and the paucity of cell adhesion data, the evolutionary relationships of the protist cell adhesion systems and components are still unclear.

2 Approaches and Findings

There have been three major approaches for identification of cell adhesion molecules. The biochemical approach relies on design of a specific cell adhesion assay followed by purification and characterization of the cell adhesion molecules. Immunochemical approaches depend on elicitation of monospecific antibodies (usually monoclonal) that block adhesion, followed by purification and characterization of the antigen. Corresponding genes have been cloned by sequence prediction, antigen expression from lambda gt11, or mutant complementation. More recently, analysis of mutants in development has come first, followed by isolation and sequencing of complementing genes, and realization that the gene encodes a cell adhesion protein. This last approach has been especially fruitful in *Caenorhabditis elegans* and *Drosophila melanogaster*.

These strategies illustrate similarities, analogies, and homologies among diverse cell adhesion systems. Most cell adhesion molecules are mosaic glycoproteins, composed of domains from a small number of gene families (Chothia 1994; Engel et al. 1994; Doolittle 1995). Almost all of them have N-terminal secretion signals. There are only a few modes of membrane attachment reported so far (type I transmembrane proteins and GPI-anchored proteins are most common). The cytoplasmic regions of the type I proteins usually have a signal transduction and/or cytoskeletal organizing function. In contrast, there is no solid evidence for any direct cytoplasmic consequence of adhesion through GPI anchored proteins. Where adhesion is Ca^{2+} dependent, the ion often affects structure and orientation of tandem, repeated domains in the cell adhesion molecules. Other similarities will emerge in the course of the chapter.

The descriptions of cell adhesion systems will start among the protists, and proceed "up the tree", (Fig. 1). Most of this chapter will concentrate on cell adhesion systems in individual species that are workhorses for cell adhesion research. The vast literature of research in vertebrate animals will not be covered, but forms a background for much of the cell and molecular biology. Following this introduction is a partial catalog describing cell adhesion systems in a variety of model organisms. The final section summarizes similarities and differences. Throughout the chapter, I have tried to reference review articles and recent findings, rather than the more "classic" primary papers. The result is a discussion of cell adhesion systems and components that is broad rather than deep.

3 Protista

3.1 Trypanosoma cruzi

Among the unicellular protists, we know most about parasite-host adhesion in the trypanosomal parasites. These organisms are kinetoplastids, and belong to an ancient grouping that includes *Euglena*. In *Trypanosoma cruzi*, penetrin, a 60-kDa non-glycosylated heparan sulfate-binding protein, has been implicated in adherence and invasion (Ortega-Barria and Pereira 1991; Herrera et al. 1994). Penetrin is an example of a protein that mediates cell-cell adhesion by binding to components of the extracellular matrix (ECM). Expression of penetrin in *E. coli* allows the bacteria to bind to and invade fibroblasts. A GPI-anchored *trans*-sialidase has been implicated in adhesion, but now appears to mediate a host inflammatory response that promotes invasion (Chuenkova and Pereira 1995). Such a dependence of invasion on host response echoes epithelial invasion by *Salmonella*. The bacteria induce an endocytic response in the host by activation of a signal transduction pathway (Cross and Takle 1993; Kaniga et al. 1995).

3.2 Cellular slime molds

These are central organisms for the study of developmental regulation of cell-cell adhesion. These organisms have a "transitional" life-style, with unicelluar and multicellular forms. The transitions from unicellularity to multicellularity depend on specific cell adhesion molecules whose characteristics and sequences are now known. Many of the characteristics of these molecules may be considered "prototypical" of the adhesion systems in the higher eukaryotes.

The life cycle of organisms such as *Dictyostelium discoideum* includes starvation-induced aggregation, signaled by pulses of extracellular cAMP, followed by migration and differentiation of the aggregate. The life cycle is described in detail in many textbooks and in several excellent reviews (Bonner 1982).

The role of most of the slime mold cell-cell adhesion proteins was elucidated by analyses of antigens identified by antiadhesive antibodies. Several adhesion systems act sequentially in aggregation of the amebae. Initial adhesion of the

D. discoideum ameba is EDTA-sensitive and mediated through an adhesion system called contact site B (csB). csB is expressed when cells are grown to high density or within a few hours of starvation. CsB adhesion appears to be localized to lateral surfaces of amebae, rather than leading or trailing edges. Two glycoproteins have been implicated: a 24-kDa protein and a 126-kDa protein. Antibodies or FAbs against gp24 block adhesion (Brar and Siu 1993). The 24-kDa glycoprotein is encoded by two closely linked genes whose products are 85% identical (Loomis and Fuller 1990). These genes are expressed early in development (timed as initiating with starvation), and expression declines after 8–12 h. gp24 has an unusual structure, having no obvious N-terminal secretion signal, but a 14-residue hydrophobic sequence from position 22 to 35. The presence of five Pro residues in this segment of gp24 would prevent α helix formation, and so this cannot be a canonical transmembrane domain. Other cell adhesion proteins have similar sequences, including algal-CAM and sea urchin sperm bindin, which may be fusogenic (see below). Mature gp24 is glycosylated; it is larger than the in vitro translation product and binds to Concanavalin A. Purified gp24 binds to *D. discoideum* cells in a Ca^{2+}-dependent manner, and gp24 itself binds Ca^{2+}. gp126 has been identified by blocking antibodies, but has not yet been further characterized. It is tempting to speculate that it is a membrane-associated receptor for gp24.

After several hours of starvation and adhesion, another cell adhesion system is expressed, known as contact site A (csA). CsA is expressed primarily on the leading edges of cells and filipodia (Choi and Siu 1987). Expression of csA depends on at least two inducers: cAMP pulses and cell-cell contact (Desbarats et al. 1994). Specifically, cohesion of ameba through csB is necessary, but not sufficient, for induction of csA, because disruption of cell contacts with EDTA or carnitine blocks expression, even in cells pulsed with cAMP. Prevention of contact by a combination of low cell density and fast shaking also inhibits induction. That csA is induced by cAMP pulses in high density cultures shaken at high speed implies that transient cell-cell contacts may be sufficient to induce the components, if the cAMP signal is present.

Adhesion through csA is through homophilic interaction: csA molecules on apposing cells interact with each other (Siu et al. 1987). Therefore, adhesion through this system may depend on expression of a single gene prodcut. This gene was originally identified in a search for mutants that were developmentally competent but lacked an epitope that reacted with a specific antibody. Mutants defective in csA form adhesions and develop under conditions of reduced shear, but have defective adhesion under mechanical stress (Harloff et al. 1989; Brar and Siu 1993). The csA gene encodes gp80, a molecule with many features common to other cell adhesion molecules (Noegel et al. 1985, 1986). gp80 has an N-terminal secretion signal, and binding activity is mediated through the N-terminal portion of the mature protein (Fig. 2). Siu and colleagues have argued that the N-terminal region contains three structures similar to Ig domains, and have identified a haptenic peptide within the first putative domain (Kamboj et al. 1989; Siu and Kamboj 1990). Cell surface anchorage is mediated by a glycosyl phosphatidylinositol anchor (GPI; Barth et al. 1994; Fig. 2). This anchor stabilizes the protein relative to an engineered version of the protein with a transmembrane domain. The C-terminal region of mature gp80 is extremely rich in the amino acids Ser, Thr, and Pro. Together, these residues constitute 42 of the C-terminal 60 residues (Noegel et al. 1986). Such regions are likely to be in extended conformation, and may serve to elevate the binding site off the cell surface (Jentoft 1990). Similar features will be mentioned in cell adhesion proteins in several other systems.

gp80 is both N- and O-glycosylated (Yoshida et al. 1993), and the carbohydrate is important for the adhesion reaction. Mannosyl residues were proposed as essential in 1985 (Hirano et al. 1985), and gp80 was identified as the major affected protein in *modB*, a glycoprotein-processing mutant with reduced adhesion (Gerisch et al. 1985). As is the case for many other adhesion systems,

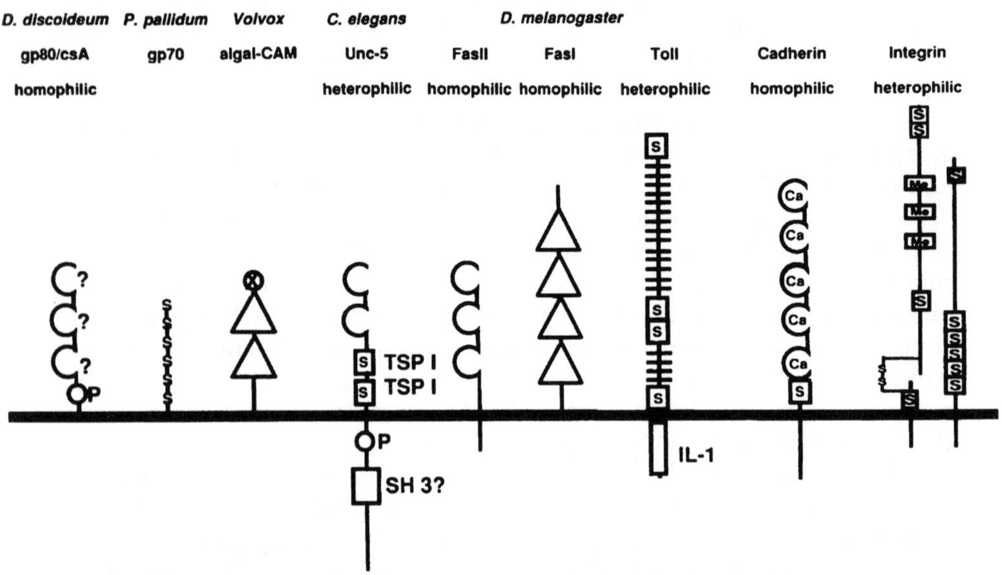

Fig. 2. Some cell adhesion molecules from the menagerie. The models are labeled with species, protein name, and mode of binding when it is known. The plasma membrane is shown as a *thick horizontal line*, and various domains are shown as *polygons* or *curves*. Vertical height of the protein is approximately to scale. GPI-anchored proteins end at the membrane and type I transmembrane proteins are shown with cytoplasmic regions. Pro-rich domains are marked *P* and Cys-rich domains are *boxed* and marked *S*. Note that such domains are not necessarily related except for their content of these residues. Disulfide bonds are *not boxed*. Ig-like domains are shown as *C-shaped*, with the gp80 structures marked *?*, and cadherin domains are marked *Ca*. Other Ca^{2+}-binding structures are similarly labeled. Metal-binding sequences in integrin are marked *Me; S* sushi domains are shown in gp70. *Triangles* denote fasI-like domains, and the *crossed* domain in algal-CAM represents extensin-like sequences. Other labels: *TSP I* thrombospondin I domains; *SH3*, *SH3*-like domain; *IL-1* cytoplasmic domain like that of the interleukin-1 receptor

we do not yet know the molecular role of the carbohydrate. gp80 saccharides whose structure has been deduced are complex-type N-linked, along with the O-linked disaccharide Galβ1-3galNAc- (Yoshida 1987; Yoshida et al. 1993).

Other cell adhesion systems are active later in *Dictyostelium* development, including contact site C, a system dependent on Mg^{2+} (Fontana 1993). At least one other EDTA-insensitive system appears around 10 h of development, and may involve a 150-kDa glycoprotein (Gao et al. 1992).

Thus, the portrait from slime mold is one of multiple cell adhesion systems, with sequential induction and activation. That expression of csA depends on function of csB implies that an inducing signal may be transduced by the adhesion system itself. On the other hand, csA null mutants develop normally under at least some conditions, so there can be no essential signal mediated by gp80/csB itself. This finding is typical of GPI-anchored cell adhesion proteins.

There are other cell adhesion systems active in other stages of the *D. discoideum* life cycle. An antigen called gp150 has been implicated in sorting of prestalk cells from prespore cells. A GPI-anchored antigen, gp138, has been implicated in mating. It is expressed at the time that mating competence develops, and is present in heterothallic and bisexual strains of *D. discoideum* (Aiba et al. 1993). The ORF encoding this protein has features reminiscent of cell adhesion proteins, including multiple N-glycosylation sites and a C-terminal region rich in Pro, Ser, and Thr (Fang et al. 1993). A definitive role in cell adhesion has not been reported.

The galactose-specific lectin Discoidin I was initially implicated in cell-cell adhesion, on the basis of its time of expression, carbohydrate-binding ability, and the observation that discoidin-binding proteins can disrupt aggregation (Breuer and Siu 1981). However, discoidin mediates adhesion to substrate glycans to promote spreading, thereby facilitating motility (Springer et al. 1984). It binds to the surface of amebae through a region including an RGD sequence. Discoidin is therefore a prototype for lectins that mediate cell-substrate adhesion.

In the cellular slime mold *Polysphondylium pallidum*, antibodies blocking aggregation identified a glycoprotein of about 70 kDa (Fig. 2; Steinemann et al. 1979; Toda et al. 1984). This antigen is a 320-residue glycoprotein that is GPI-anchored and N-glycosylated at all six sites with the consensus N-glycosylation sequence Asn-Xaa-Ser/Thr (Saito and Ochai 1993). It appears not to be O-glycosylated (Steinemann et al. 1979; Manabe et al. 1994). The protein contains 36 Cys residues, and at least 30 of them form disulfide bonds (Saito et al. 1993a, 1994). Twenty four of these residues are in "sushi" domains, structures of 14 to 21 sequential residues with two disulfides, where the first Cys is bonded to the third, and the second bonded to the fourth. Such structures have not been previously reported in cell adhesion molecules, but are found in several components of the complement cascade, and may, in fact, be common extracellular protein-protein interaction motifs. The mode of adhesion is unknown, but fucose inhibits a blocking mAb, implying a possible role for carbohydrate (Toda et al. 1984).

To summarize briefly, cell-cell adhesion in slime molds is characterized by multiple systems, whose expression is developmentally regulated. EDTA has proven useful in functional separation of cell adhesion systems. Several glycoproteins involved in cell adhesion are anchored to the outer leaflet of the plasma membrane by GPI anchors, and at least one cell adhesion protein is homophilic, mediating cell-cell contact through self-association. In several cases, carbohydrate is suspected as a determinant for adhesion, but a definitive role has not been found. The relatively well-characterized lectin discoidin mediates cell-substrate interactions, a mechanism that is common in other organisms.

4 Higher Eukaryotes

rRNA sequence alignments imply that the major multicellular kingdoms diverged from the protista within a relatively short time (Patterson and Sogin 1992; Conway 1994). Thus, the animals, plants, and two other less familiar goupings (alveloates and stramenophiles) diverged around a billion years ago. The rRNA sequences clearly place the fungi along the evolutionary path leading to animals, a surprising result (Wainwright et al. 1993). This possibility is consistent with the finding that there are cell adhesion proteins that are members of gene superfamilies common to animals and fungi, but not to plants. However, the commonality of fungal and animal sequences may be an artifact of our ignorance of cell adhesion proteins in the other kingdoms.

5 An Alveolate: Plasmodium

In two stages of the malaria life cylce, we know something about the molecules that mediate adhesion to host cells. An erythrocyte-binding protein called Duffy antigen-binding protein (DABP) or erythrocyte-binding protein-175 (EBA-175) mediates adhesion to erythrocytes and invasion. These adhesion proteins show different ligand specificities in different species and strains, but possess a common sequence called a Duffy antigen-binding domain (DBL), after the first ligand discovered (Chitnis and Miller 1994; Borst et al. 1995). These adhesins are type I transmembrane proteins with extracellular domains of about 1000 residues that include two or three Cys-rich DBL domains. In *P. falciparum* EBA-175, a DBL domain of about 120 residues is sufficient for binding to human glycophorin A in a manner dependent on sialic acid and peptide sequences in the gylcophorin (Adams et al. 1994; Sim et al. 1994).

A related set of plasmodial genes (*var*) encode adhesion proteins expressed on the erythrocyte surface. These proteins allow infected erythrocytes to adhere in capillaries, and escape destruction in the spleen (Borst et al. 1995). These 200–350-kDa proteins mediate adhesion to unrelated endothelial surface proteins including thrombospondin, CD36, ICAM-1, VACM-1, and ELAM-1

(Baruch et al. 1995; Smith et al. 1995; Su et al. 1995). Frequent expression switching between 50–150 *var* genes allows antigenic escape as well as diversification of binding specificities. These proteins have a conserved N-terminal sequence with one DBL repeat and a similarly sized Cys-rich domain of related sequence. One to three more copies of the DBL domain follow, with nonconserved spacer sequences in between. In various *var* gene products the transmembrane domains are followed by cytoplasmic domains of 16–500 residues. The longer cytoplasmic domains are highly acidic. That proteins with similar domain structure should recognize such a variety of ligands implies that the binding sites have different structures, but are built on a "scaffold" of constant structure. This is exactly the case for members of the Ig superfamily, and especially the antibodies, where binding sites are located in hypervariable loops at the ends of β barrels of constant structure and conserved sequence (Williams and Barclay 1988).

6 Plants

In the plants and the fungi, there is an additional complexity to cell adhesion; the presence of the cell wall. Initial recognition may depend on interaction of wall components, rather than membrane proteins. Wall components may be considered analogous in some way to the extracellular matrix of animal cells, and so initial cell adhesion may resemble cell-substrate adhesion rather than direct cell-cell interactions. Thus, initial recognition is analogous to interaction of the outer layers of the egg and sperm in fertilization in animals. Other cell adhesion systems mediate membrane-membrane interactions.

6.1 Chlamydomonas

The mating of the unicellular green alga *Chlamydomonas* depends on specific cell adhesion systems. This genus is proposed to have split from the evolutionary branch leading to the higher plants at a time close to the divergence of the Precambrian crown, and so may incorporate elements common to the protists and the plants (Goodenough 1991). *C. reinhardtii* and *C. eugametos* diverged soon after the radiation from the higher plant line, so that characteristics common to these species may be common to organisms near the origin of the Precambrian crown. In both species, initial adhesion is through glycoproteins that are localized to the flagella. *mt+* and *mt–* cells adhere through a species- and mating-type-specific interaction that requires no known ionic cofactors. The adhesins are large proline-rich molecules related in sequence and structure to cell wall components. The isolated molecules appear as extended structures 225 nm long, with a hook structure at the flagella-proximal end and a head or knob at the distal end (Goodenough 1991). Adhesion is protease-sensitive, and treatments that destroy the head also disrupt the adhesion. Side-to-side adhesion of the shafts is found early in the interaction, and resembles molecular interactions in self-assembly of cell wall proteins (Adair and Snell 1990). Adhesion results in cross-linking of the agglutinins, followed by movement of the adhesive region toward the tips of the flagella, and apposition along the entire length ("tipping"). The cross-linking triggers a second messenger cascade, probably through induction of adenyl cyclase (Saito et al. 1993b). Activation of the cyclase or addition of dibutyryl-cAMP is sufficient to induce a variety of mating behaviors, including mobilization of cryptic adhesins (Tomson et al. 1990). Thus, adhesion itself increases the cell surface concentration of active adhesins, a strategy that should increase the probability of forming productive interactions between mating cells. Elevation of cAMP levels also results in lysis of the cell wall (Adair and Snell 1990) and activation of mating structures (Goodenough 1991). These structures are fusion-competent areas of the cell membranes that are rapidly assembled and activated. This actin-driven formation of "mating structures" in a specific apical region of each mating cell resembles the acrosome reaction of sperm. A molecule of the *mt*+ mating structure that appears to mediate specific adhesion of the + and − mating structures has been recently cloned and sequenced. Mutants in this gene adhere normally through the flagella, but the mating structures fail to adhere. The corresponding ORF encodes a type I transmembrane protein that could be a cell adhesion molecule (U.U. Goodenough, J. Woessner, pers. comm.).

The presence of two cell adhesion systems active in mating of *Chlamydomonas* illustrates several general observations. As in slime mold aggregation and leukocyte adhesion to endothelia, cell interactions are mediated by multiple systems acting sequentially (Springer 1990; Lawrence et al. 1995). In mating interactions, there is often initial contact through peripheral adhesion molecules and a later, more intimate, step involving specific adhesion of membranes prior to fusion.

6.2 *Volvox*

At least one cell adhesion family is present in algae and animals. A monoclonal antibody that disrupts embryogenesis in *Volvox* identifies a presumptive cell adhesion glycoprotein (called algal-CAM) present on cell surfaces (Huber and Sumper 1994). The nine exons encode multiple isoforms by alternative splicing, and different forms are present at different stages of development. The sequenced isoforms contain a Pro-rich N-terminal with sequence motifs characteristic of extensin, a cell wall structural protein of plants and alga. A region with two repeats of a 150-residue sequence follows. This repeat is called FasI, after a fasciclin of *Drosophila* (see below). In one form of algal-CAM, a hydrophobic C-terminal sequence implies the existence of a GPI-anchored form, a structure not well characterized in plants. In other forms, there is no canonical membrane association domain, but there is a short hydrophobic sequence near the C-terminus, as in *D. discoideum* gp24 and sperm bindin.

6.3 Higher Plants

Little is known about specific cell adhesion in higher plants. Among genes necessary for self-incompatibility and fertilization in *Brassica* and *Arabidopsis* is a family encoding the S proteins. These proteins have several common regions, and a structure reminiscent of cell adhesion proteins: they share extracellular sequences that could form compact domains, a Cys-rich region, a transmembrane domain, and in many members, a cytoplasmic Ser/Thr kinase domain (Pruitt et al. 1993). Similar molecular designs are also found in the immunoglobulin superfamily in animals (Williams and Barclay 1988). As in the Ig superfamily, some members of the S protein gene family have only extracellular domains, and may be localized to the cell wall, as in the yeasts (see below). Although these molecules are required for specificity of pollination, there is as yet no evidence that this gene family directly mediates cell adhesion.

7 Fungi

7.1 Saccharomyces cerevisiae

In baker's yeast, recognition of the haploid mating types **a** and α is mediated through interacting glycoproteins, **a**-agglutinin and α-agglutinin, respectively (Lipke and Kurjan 1992). Their structures are summarized in Fig. 3. Direct interaction of these agglutinins has been demonstrated by immunological and biochemical means (Cappellaro et al. 1991). In other budding yeasts, the sexual agglutinins appear to be homologous and similar in structure (Lipke and Kurjan 1992).

α-Agglutinin is the product of the *AGα1* gene, and is expressed only by mating type α cells (Fig. 3). The glycoprotein consists of four regions: an N-terminal secretion signal, a 307-residue recognition structure, a 300-residue highly glycosylated Ser/Thr-rich stalk region, and a C-terminal signal for addition of a GPI anchor (Wojciechowicz et al. 1993). The recognition domain consists of three domains that have sequence and structural similarity to immunoglobulin variable domains. Two disulfide bonds stabilize the domains, and there are two free sulfhydryl groups (Chen et al. 1995). Binding to the ligand **a**-agglutinin requires His^{292} in the third domain as well as residues close to it in domain 3 and unidentified residues in domain 1 and/or 2 (Cappellaro et al. 1991; Chen et al. 1995; de Nobel et al., 1996). A three-dimensional model of domain 3 shows the essential residues clustered at one end of the *β* barrel of the domain, opposite from the binding region of the immunoglobulins, but similar in location to the binding residues of the human growth hormone receptor (Bass et al. 1991). α-Agglutinin binds to its ligand **a**-agglutinin with at least two affinities, a weak one of unknown affinity and a cold-sensitive binding of 10^9 L M^{-1} (Lipke et al. 1987). There is no evidence that binding requires metal ions.

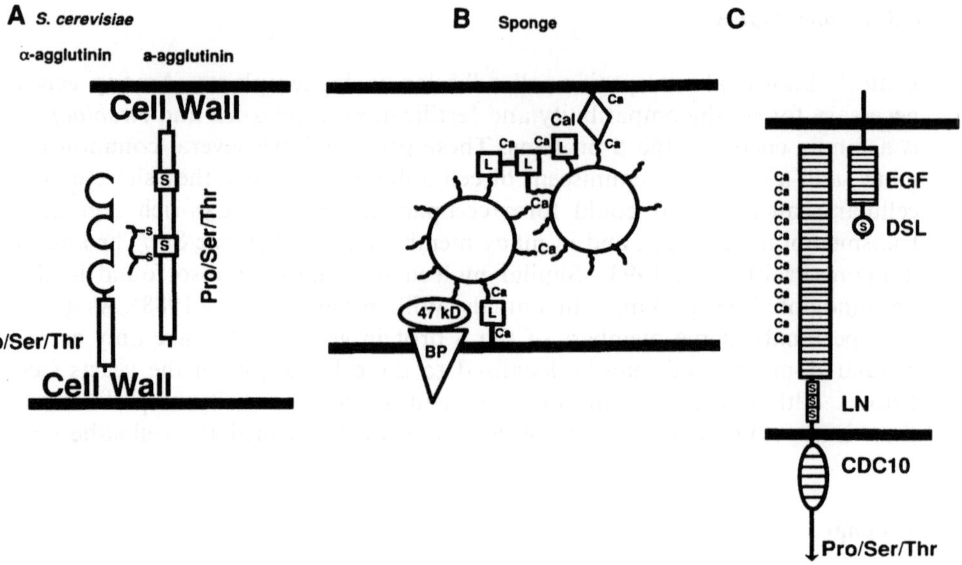

Fig. 3A–C. Three heterophilic cell adhesion systems. Symbols as in Fig. 2, or labeled. **A** *S. cerevisiae* sexual agglutinins. **B** Some of the possible interactions for a sponge AF "chimera" of the studied species. *L* Lectin; *Cal* calpactin; *BP* baseplate or cellular receptor; *47 kD* AF receptor. This model is not to the same scale as the others in Figs. 2 and 3. **C** *D. melanogaster* Delta and Notch proteins

The Pro/Ser/Thr-rich stalk of α-agglutinin contains 7 N-glycosylation sites, and 125 O-glycosylation sites, all of which may be used (Wojciechowicz et al. 1993; Chen et al. 1995). This region has been predicted to constitute a 50–70 nm stalk that holds the binding site out from the surface of the cell wall (Cappellaro et al. 1994; Chen et al. 1995). The GPI anchor is present only during biogenesis of the agglutinin; it is subsequently processed to mediate covalent anchorage to the cell wall polysaccharide (de Nobel and Lipke 1994; Lu et al. 1995)

a-Agglutinin is the product of two genes, *AGa1* and *AGa2*. The former encodes a 725-residue cell wall anchorage polypeptide that is highly O-glycosylated and contains Cys-rich domains (Lipke and Kurjan 1992). Like α-agglutinin, it has sequences predicted to function as a secretion signal and a GPI anchorage signal. It is also covalently bound to the cell wall matrix. Remarkably, it is expressed in both haploid mating types (Lipke and Kurjan 1992; de Nobel et al. 1995). This finding is surprising, but may be explicable, because this Pro/Ser/Thr-rich glycopeptide is prototypical of a large class of cell wall glycoproteins in yeast, and could be a minor wall structural protein (Teunissen et al. 1993; de Nobel and Lipke 1994; Klis 1994; van der Vaart et al. 1995). Therefore, as in *Chlamydomonas* and *Volvox* adhesins (Goodenough 1991; Huber and Sumper 1994), yeast agglutinins contain regions homologous to cell wall proteins.

At least two of the Cys residues in the **a**-agglutinin anchorage subunit Agalp are in disulfide linkage with the binding subunit, the product of the *AGa2* gene (Fig. 3). This glycopeptide (69 residues after signal cleavage) carries 30% by weight O-linked carbohydrate, and activity is destroyed by α-mannosidase or periodate treatment, implying an essential role for the carbohydrate (Cappellaro et al. 1994). This peptide is expressed only in mating-type **a** cells. The C-terminal of the peptide blocks adhesion, and an unglycosylated synthetic peptide was also inhibitory, although it was 25-fold less potent than the intact binding subunit (Cappellaro et al. 1994). These results suggest that the glycosylation may serve to maintain the structure of the ligand determinant, rather than be a determinant itself (Wyss et al. 1995). Such a role would be consistent with the inactivity of mannosides or yeast oligosaccharides as inhibitory haptens.

Physiological Role of the Agglutinins. When juxtaposed on solid medium, mutants with any of the agglutinin genes deleted mate at an efficiency approaching that of wild-type cells (de Nobel et al. 1995). This result argues that the agglutinins have no essential function in mating, and do not transmit essential mating signals. All such signals are mediated by the peptide sex pheromones, which induce mating-specific functions and apparently induce fusion behavior when appropriate (Kurjan 1993). Pheromone effects include induction of mating-specific gene products, mate selection (response to the potential partner secreting the most pheromone), cell wall remodeling, and induction of fusion competence.

In contrast, the agglutinins are important in mating in liquid phase. Matings of non-agglutinable mutants are 10^{-4} to 10^{-6} less efficient than those of isogenic agglutinable strains (de Nobel et al. 1995). This sterility persists under conditions that promote cell contact: mating in an undisturbed sediment or within a centrifuged cell pellet (Lipke and Kurjan, unpubl.). Thus, cell-cell contact is not sufficient to promote mating in liquid media.

This paradox can be resolved by consideration of the response to the pheromones. Although each pheromone binds to a single receptor on its respective target cell, there are at least two signal transduction pathways induced in each mating type (Baffi et al. 1984, 1985; Jackson et al. 1991). At least one of these pathways, that leading to morphogenesis and cell fusion, responds only to prolonged (90-min) pheromone binding to receptor at pheromone concentrations 100-fold higher than the minimal effective dose (Baffi et al. 1984). These pheromone concentrations are substantially greater than that of the bulk solution, especially for the pheromone a-factor, which is farnesylated and relatively insoluble (Kuchler et al. 1993; Kurjan 1993). A model to explain this result is shown in Fig. 4. In a mating mixture, cells are randomly oriented, but respond by morphogenesis and growth toward the pheromone source of greatest potency. Since secretion in yeast is polarized (Jackson and Hartwell 1990; Kuchler et al. 1993), the cells of each mating type not only grow from a single point on the cell surface, but also secrete pheromone, receptor for the

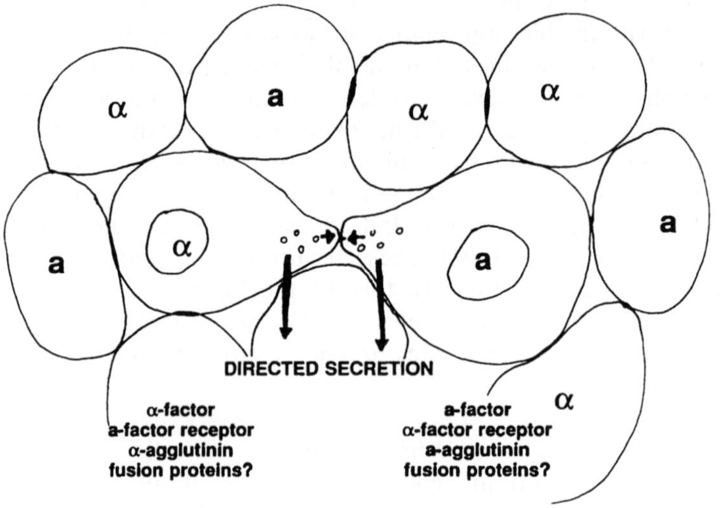

Fig. 4. Relationship of cell adhesion to signal transduction in *S. cerevisiae*. Mating cells adhere most strongly at the tips, which are sites of polarized secretion

conjugate pheromone, and agglutinin at the same site. Thus, response to pheromone turns on a positive feedback loop that leads to increased local secretion of each extracellular component of the signaling pathway, reinforcing and prolonging the pheromone response, and reinforcing the adhesion as well. Such recruitment of cell adhesion molecules was mentioned for *Chlamydomonas* (Tomson et al. 1990), and is a common feature of adhesions activated during differentiation in mammalian cells as well (Springer 1990). The strengthened adhesion of the cellular secretion domains ensures that the local pheromone concentration is substantially higher than in bulk solution, and therefore facilitates all pheromone responses. In two-dimensional matings (on agar or filters), pheromone diffusion is limited to the boundary water layer, and local pheromone concentrations are high enough to induce all mating responses, even in the absence of agglutination. Thus, the agglutinins are "essential" in a three-dimensional world, but superfluous in a two-dimensional situation.

Other Cell Adhesion Systems. There is as yet no reported characterization of the adhesion system mediating membrane fusion in yeast mating, but such a system must exist. Some yeasts also have a nonspecific, Ca^{2+}-dependent cell adhesion ability called flocculation. The molecular basis for cell-to-cell binding is not known, although several necessary genes have been identified, including *FLO1*, which encodes a cell surface glycoprotein (Teunissen et al. 1993). This protein is Ser/Thr-rich and GPI-anchored, and is presumed to be a cell wall anchorage subunit analogous to the **a**-agglutinin anchorage subunit. Flocculation can be partially inhibited by hundred millimolar concentrations of α-mannosides, and so shows a carbohydrate-binding component, analogous to that in sponges and slime mold. Flocculation is reminiscent of polysaccharide-based coadhesion of bacteria, and may also serve to form cohesive colonies that provide advantageous microenvironments for growth (Schilling and Bowen 1992).

7.2 Candida

In the yeast *C. albicans*, several sets of potential cell adhesion molecules have been demonstrated. This organism is parasitic or commensal in mammalian epithelia, and has evolved a large variety of mechanisms for adherence to epithelial cells of mammals. There are components that bind complement (Calderone 1993), receptors for RGD sequences in complement or fibronectin (Bendel and Hostetter 1993; Pendrak and Klotz 1995), lectin-like activities (McCourtie and Douglas 1985; Ollert et al. 1993), as well as the nonspecific hydrophobic interactions of the germ tube (pseudohyphal) form (Glee et al. 1995). It is not known if these adhesive capabilities were independently evolved or acquired from mammalian cells by horizontal gene transfer after development of the commensal relationship (Sprague 1991).

Of the adhesive molecules, none is yet well characterized at the sequence level, but there are two proteins of interest. *C. albicans* and the *C. tropicalis* express a protein epitope for monoclonal antibodies specific to $\beta 1$ subunits of integrins (Bendel and Hostetter 1993). Such epitopes are not found in *S. cerevisiae* (Marcantonio and Hynes 1988). These "integrin analogs" (Bendel and Hostetter 1993) may be the receptors for the RGD proteins. The absence of integrin-like sequences from *S. cerevisiae* is compatible with loss of the gene from that species, or with acquisition of the molecule late in evolution of *Candida*. Secondly, an apparent homolog of *S. cerevisiae* α-agglutinin has been identified in *C. albicans* and related species. In *C. albicans*, the mRNA product of this gene, *ALS1*, is expressed upon generation of germ tubes, which are substantially more adhesive than the yeast form. However, an adhesive function of *ALS1* has not been shown. *ALS1* does not encode a mating agglutinin, because *Candida* species do not mate. That Ig-like domains are present in animal and fungal cell adhesion glycoproteins is a strong argument that cell recognition function was established before the kingdoms diverged.

8 Metazoa

The animal kingdom shares key features that include multicellularity, homeotic genes regulating development (Gamulin et al. 1994; Carroll 1995), invagination of the blastula to form a tubular body plan (Barnes 1980), and extracellular matrices containing collagens. rRNA sequence analysis confirms the morphological conclusion that the Metazoa are monophyletic in the sense that they share a common ancestor not shared by members of any other kingdom (Fig. 1; Wainwright et al. 1993). It is not surprising, therefore, that cell adhesion systems central to the development of multicellularity and the morphogenic process should share structural and functional roles throughout the kingdom (Chothia 1994).

8.1 Sponges

Cell adhesion and cell adhesion molecules were first identified in sponges (Wilson 1907; Margoliash et al. 1965; Cauldwell et al. 1973) and much is known about cohesion of cells in adult sponges. Such adhesion is often species-specific, being mediated through proteoglycan "aggregation factors" (AF) that are unique to each species. Such associations are analogous to cell adhesion to extracellular matrix in more complex animals, in that many cells remain motile within the matrix (Müller et al. 1982), and that metal ions (especially Ca^{2+}) mediate adhesion. The reaggregation and sorting behavior of isolated sponge cells is also analogous to that of embryonic cells from vertebrates. Cells of different sponge species cohere, but often sort from each other upon subsequent migration (Holtfreter 1943; Moscona 1957; Cauldwell et al. 1973).

Sponges, including Geodia *cydonium*, can be dissociated following soaking in seawater free of Ca^{2+} and Mg^{2+}. In reaggregation of the individual cells there are two cell adhesion systems, a collagen-based initial formation of small aggregates (Müller et al. 1982; Diehl-Seifert et al. 1985a) and the aggregation factor-dependent formation of large aggregates (Müller et al. 1982). Aggregation factors (AF) are large complexes ($>10^7$ Da) that self-aggregate, and include proteoglycans and associated proteins that are cellular effectors (Fig. 3B). Thus, the AFs are like extracellular matrix in mammals, in that they are associations of many proteins and activities (Scott 1992). AFs bind to cell surface receptors (Müller et al. 1976; Kuhns et al. 1980; Schröder et al. 1989).

AFs are visible in the electron microscope as branched structures; the *G. cydonium* AF has a "sunburst" structure, with the radial arms mediating AF-cell interaction; in *Microciona prolifera*, the AF is the "brush-like". In several species, the glycan appears to be composed of repeating units that have been implicated as critical for aggregation (Misevic and Burger 1986, 1993; Parish et al. 1991; Spillman et al. 1995). In *G. cydonium*, AF contains a gal-specific lectin (Diehl-Seifert et al. 1985b). The lectin is self-aggregating, and also binds to AF and cells. It could thus form a substratum for cell attachment to AF or other ground substance. Activity of the lectin is regulated by Ca^{2+} at concentrations typical of that in seawater. As with other instances of carbohydrate-binding components in cell adhesion systems, we do not have a clear idea of the physiological significance of the lectin. A sialyl transferase is also associated with AF, and sialylation of the receptor is reported to be necessary for activity (Müller et al. 1977).

Other components associated with *Geodia* AF include a 47-kDa cell-binding component (Gramzow et al. 1986). Antibodies against this protein inhibit AF-dependent cell aggregation, and the protein interacts with the cellular aggregation receptor protein. Cell binding is species-specific, but is not regulated by Ca^{2+}. Another cell-AF interaction is strongly Ca^{2+}-dependent: the Ca^{2+}-binding protein calpactin is found in AF, and mediates interaction with cell surface phospholipids (Robitzki et al. 1990). This protein is 83% identical to the sequence of calpactin II from rat. Antibodies against this protein prevent sponge cell aggregation. The sponge protein binds 2 mol Ca^{2+} per mol protein, with a K_d between 500 and 800 μM. Phospholipid binding is regulated by Ca^{2+} in a hysteretic manner, with half maximal lipid binding seen at Ca^{2+} concentrations around 500 μM, but lipid dissociation promoted at Ca^{2+} concentrations between 5 and 50 μM. (The Ca^{2+} concentration in the interstitial space of sponges and in seawater is about 10 mM.)

Burger's group is carefully characterizing AF from *M. prolifera*. AF is about 60% carbohydrate and 40% protein (Misevic and Burger 1986). Several extracellular matrix proteins are associated with AF, including glycoproteins of 68 and 210 kDa (Varner et al. 1988). AF can be proteolytically cleaved to produce 10-kDa fragments equivalent to 60% of the mass of the AF. These fragments bind *M. prolifera* cells with low affinity (K_d = 3.4 10^{-7} M), and strong binding of intact AF is a result of multipoint attachment. Generation of high affinity by multivalent binding has also been observed in agglutination of the yeast *Hansenula wingei* (Lipke and Kurjan 1992).

In a classic study, Jumblatt et al. (1980) demonstrated that cell-cell adhesion is mediated by specific Ca^{2+}-dependent carbohydrate-carbohydrate interactions of the AF. Cell surface receptors also bind carbohydrate determinants, but as in *G. cydonium*, the cell-AF interactions are not dependent on Ca^{2+}. As an approach to the determination of the critical structures, anticarbohydrate monoclonal antibodies that block aggregation have been produced. The structures of the epitopes reveal that at least two acidic sugar fragments are involved in adhesion: a pyruvate diacetal structure pyr2–4,6Galβ1–4GlcNAcβ1–3Fuc-, which has a free carboxyl group, and SO_4-3GlcNAcβ1-3Fuc-. The acidic groups could bind Ca^{2+}, and a stereospecific binding could mediate specificity of the interaction in a manner that would only allow species-specific aggregation, as has been demonstrated for larger glycans (Misevic and Burger 1993).

Müller's group and Burger's have argued that binding of AF to receptor causes Ca^{2+} efflux from cells, which would raise the Ca^{2+} concentration near the cell surface, facilitating primary and secondary aggregate formation (Müller et al. 1987; Spillman et al. 1995). Thus, initial adhesion stimulates further adhesion, and an adhesive "cascade" based on a positive feedback loop is also present in this system. The increase in Ca^{2+} would activate collagen assembly factor (primary adhesion), calpactin (a secondary or tertiary interaction), and lectin. The lectin is mitogenic, so adhesion would also promote growth of the organism (Müller et al. 1987). The efflux of Ca^{2+} is triggered through release of polyphosphoinositides and diacyl glycerol release, leading to activation of protein kinase C (Weissman et al. 1988; Schröder et al. 1989) as well as Ca^{2+} mobilization. Thus, in sponge, there is a clear instance of cell adhesion triggering a second messenger cascade.

With the advent of PCR techniques, cloning and sequencing of genes from sponges has become possible. One of the first interesting results is the identification of a tyrosine kinase homolog from *G. cydonium*. The predicted extracellular portion of this ORF has sequence motifs similar to immunoglobulin superfamily members. However, the extracellular sequence is too short to encode a classical Ig domain, as it includes only 59 residues (Schäcke et al. 1994).

8.2 Cnidaria: Hydra

Hydra is a member of one of the phyla with radial symmetry, and is diploblastic, having only two cell layers. Fibronectin, collagen type IV, and laminin are found in the mesoglea, an ECM that separates the ectoderm and endoderm of hydra. Migrating nematocytes adhere to mesoglea, and to human fibronectin. The adhesion to fibronectin is inhibited by a peptide containing the sequence RGD, a characteristic of binding of integrins to fibronectin and other ECM material. A fibronectin column retained cell surface proteins of sizes reminiscent of integrin subunits (Ziegler and Stidwell 1992). Thus, the ECM in this organism has many of the features of that in more complex animals, and adhesion to the ECM is probably mediated by similar mechanisms.

8.3 Triploblasts

The diploblastic phyla (sponges and radiata) and the triploblasts were clearly differentiated between 600 and 1000×10^6 years BP (Conway 1994). There is little information on cell adhesion in the radiata or in the oldest triploblastic phyla. Clearly, research into the genes and phenomena of cell adhesion would add much to our understanding of the development of multicellularity in the animal kingdom. Hydra and flatworms are obvious model organisms (Barnes 1980).

8.3.1 Pseudocoelatomates: Caenorhabditis elegans

The determination of cell lineage and developmental interactions in *Caenorhabditis elegans* has given great insight into the mechanisms and uses of cell adhesion molecules by "looking through the back door." Genetic and developmental approaches have identified a number of direct cellular interactions, and the rapid progress of sequencing of the genome of this nematode has shown us some of the genes involved and their modes of action (Sternberg 1993; Chothia 1994; Stern and DeVore 1994). Thus, homologs of S-type lectins

(Hirabayashi et al. 1992), cadherins (Sano et al. 1993), integrins (Marcantonio and Hynes 1988; Gettner et al. 1995), and members of the Ig superfamily (Leung-Hagesteijn et al. 1992) have been sequenced. In several cases, phenotypes of mutants in specific genes have given clues as to function. The extracellular matrix contains many recognizable proteins common to vertebrates and invertebrates (Engel et al. 1994). It is relevant to point out that these homologs are easily recognizable by sequence similarities between the *C. elegans* and vertebrate proteins, so that these cell adhesion proteins have substantially conserved sequence of key residues since their divergence at least 540 $\times 10^6$ years BP (Conway 1994).

Cell-Substrate Interactions. Two recently discovered roles for cell adhesion in *C. elegans* presage developmental roles in higher animals. UNC-5 (Fig. 2) is a cell surface protein that includes two Ig domains, two domains homologous to thrombospondin (a Cys-rich ECM protein that may bind to sulfated glycans), a transmembrane domain, and an intracellular tail that includes a Pro-Ser-rich sequence and an SH3-like domain, analogous to signal transduction domains in mammalian cells (Leung-Hagesteijn et al. 1992). Thus, the extracellular domains are typical of cell adhesion proteins, being a mosaic of several types of domains (Engel et al. 1994; Doolittle 1995). UNC-5 is essential for dorsal guidance of pioneering neurons. UNC-6, a laminin homolog, may be a ligand for UNC-5, and would be laid down as part of the basal lamina. Binding to UNC-5 might trigger a positive response in axon migration.

Another component of the basal lamina affects cell-substratum adhesion in development. *UNC-52* encodes a homolog of perlecan, a highly mosaic core protein for heparan sulfate glycosaminoglycans. *UNC-52* encodes several alternatively spliced forms of a large protein with four regions: I, the secretion signal and a short acidic sequence; then an IgC2 domain; region II, containing three Cys-rich sequences similar to those found in LDL receptor; another IgC2 domain; region III is similar to the amino-terminal region of laminin, including seven Cys-rich sequences and two large globular domains; and IV a region with 14 IgC2-like domains in tandem and a highly basic C-terminal sequence (Rogalski et al. 1993). An *unc52* mutant has defects in muscle attachment, and the sequenced allele showed a transposon insertion in the IgC2-rich region. A suggestion that alternatively spliced forms are essential follows from the observation that *mec-8* mutations, which may be splicing-defective, are synthetically lethal with *unc-52* (Lundquist and Herman 1994).

Cell-Cell Adhesion. We now know that a critical type of cell-cell interaction is homologous in nematodes, insects, and mammals. The *C. elegans* genes *Lin-12/Glp-1/Lag-2/Apx-1* are required for specification of cell fate, following a model discovered in the *Drosophila* genes *Notch/Delta/Serrate* (Fig. 3); (Artavanis-Tsakonas et al. 1995; Simpson 1995). Cell surface receptors on one cell (products of *Lin-12* or *Glp-1*) interact with ligands on the surface of apposed cells (Lag-2 or Apx-1). The Lin-Notch family consists of type I transmembrane proteins with (from the N-terminus) many repeats of an EGF-like domain (family members with 11–36 repeats are known), three repeats of a Cys-rich sequence characteristic of the gene family (LN domain), a transmembrane domain, six repeats of a sequence found in ankyrin and yeast *CDC10, SW14,* and *SW16* (genes involved in cell cycle regulation), and a Pro-Ser-Thr-rich C-terminal region (Fig. 3c). The ligands have a Cys-rich domain called DSL (*Delta Serrate Lag*; the sequence shows some similarity to EGF repeats, and the Cys residues are essential), 1 to 15 EGF domains, a transmembrane domain, and a short cytoplasmic tail that is different in each ligand. Replacement of the *C. elegans* Lag-2 intracellular domain with a β-galactosidase sequence resulted in wild-type embryos, so the intracellular domain may not be essential (Henderson et al. 1994). Binding of ligand to the Notch/Lin-12/Glp-1 receptor triggers a signal transduction event that leads to altered transcriptional patterns and determination of cell fate (Artavanis-Tsakonis et al. 1995). On cell-cell contact, receptor clustering is initiated, and the transcription factor "Suppressor of Hairless" [Su(H)] is activated and/or displaced from the Lin-12/Glp-1/Notch intracellular domain, and perhaps translocated to the nucleus. An inductive interaction mediated by this system is seen in *C. elegans* germ cell development: Lag-2 expressed by the distal tip cell interacts with Glp-1 on the germ cells of the hermaphrodite gonad to promote mitosis and proliferation of the germ line. In the absence of the *Glp-1/Lag-2* signal, the germ cells enter meiosis and produce primarily sperm cells with few oocytes (Stern and DeVore 1994). The gene *Lin-12* is essential for a fate decision as to which of two cells becomes the ventral uterine precursor (vu cell), and which the anchor cell (ac). In this system, both cells express Lin-12 and a signal inhibiting ac fate in the other cell. The two cells compete to inhibit ac fate in each other, with one of the cells eventually winning. Thus, a deletion of either cell causes the other to become ac, as do *lin12* null mutations (Greenwald 1989). In contrast, some *lin-12* mutations are homeotic and cause both cells to acquire a vu fate. A similar effect of the *lin-12* mutations is seen in development of the vulva: the primary cell acts through *lin-12* to prevent 1° fate in neighboring cells; in *lin-12* null embryos, all presumptive vulval secondary cells acquire primary fate and in *lin-12*(*d*) mutants all vulval precursor cells are secondary (Stern and DeVore 1994). Similar genetic and cell fate experiments demonstrate that *glp-1* is required for anterior-posterior and left-right axis determination in cell fate at the 8- to 12-cell stage of the embryo (Hutter and Schnabel 1994). This paper describes remarkable experiments in which cell fate was altered by micromanipulation of individual target cells so that they resided next to and presumably adhered to specific "fate-inducing" cells in the embryo. Alteration of cell fate was dependent on the presence of *Glp-1*. The results are consistent with cell-cell-contact-dependent determination of cell fate through Glp-1. The determination of cell fate through mediation of cell-cell adhesions echoes a realization that neighbor cell interactions are often required for proper differentiation and histogenesis (Vardimon et al. 1988; Harloff et al. 1989; Artavanis-Tsakonas et al. 1995).

8.4 Insects

As with other areas of development, the fruit fly has become a well-characterized model system for cell-cell and cell-substrate adhesion. As in *C. elegans*, cell adhesion molecules have been primarily identified by sequencing of genes corresponding to mutations that affect development. *Drosophila* cell adhesion proteins include members of all major vertebrate cell adhesion families, and some whose vertebrate homologs are currently being identified. A brief discussion of major classes of cell adhesion proteins follows in this order: Notch/Delta, immunoglobulin homologs, fasciclin IV, fasciclin I, leucine-rich-repeat proteins (LRRs), cadherins, and integrins. ECM molecules are not reviewed here (Murray et al. 1995).

EGF Repeat Family. The *Notch/Delta* system directly determines cell fate in mesodermal and ectodermal tissues (Corbin et al. 1991). Fate specification follows binding of Delta protein to Notch, with subsequent activation of a signaling pathway in the cell expressing Notch, as in *C. elegans* (Artavanis-Tsakonis et al. 1995). Activation of this pathway inhibits a default cell fate. Mutations in component cell adhesion proteins cause alterations in cell fate, with too many or too few cells adopting the wild-type fate. Such interactions are often pairwise: single pairs of cells signal each other to induce a specific fate in one of them and a default fate in the other, a mechanism to ensure that only a single cell differentiates to the specific fate (Ghysen et al. 1993). In other cases, a single cell signals to several others to determine fate, as in the 1° cell inhibition of 1° cell fate in its neighbors in the nematode vulva. Mutations in *Notch* result in hypertrophy and disorganization of the nervous system, eye defects, wing deformation, disorganization of egg chambers, and altered bristle patterns.

Notch and Delta can act as classic cell adhesion molecules (Fig. 3), and mixtures of cells carrying these proteins aggregate specifically. The interaction is heterologous, and neither Notch nor Delta mediates homophilic adhesion. EGF repeats 11 and 12 of Notch are necessary and sufficient for adhesion to Delta-expressing cells (Rebay et al. 1991). However, single-site mutations in EGF repeats 14, 24, 25, 27, or 29 lead to alterations of cell fate, and some of these mutations can be suppressed by specific mutations in Delta, implying Notch-Delta binding in these regions. Such mutations also change adhesive properties of the proteins, decreasing aggregate size and ability to compete for binding to wild-type proteins (Lieber et al. 1992).

Another member of the EGF family is *Crumbs,* which encodes a type I transmembrane protein with 30 EGF repeats and 4 laminin A G repeats. Mutations in this gene lead to failure to localize apical markers in epithelia, and overexpression of Crumbs protein leads to gross expansion of the apical domain (Wodarz et al. 1995). Cytoskeletal reorganization is apparent in both types of mutant, implying that the 37-residue cytoplasmic tail is critical for coordination of membrane-cytoskeletal specificity. Crumbs is not known to be a cell adhesion protein, nor is a ligand known.

Members of the Immunoglobulin Superfamily (IgSF) in Drosophila. As in vertebrates, members of the IgSF are necessary for proper neural development. In insects, neural pathways are reproducibly guided in defined pathways during development. Goodman's group has used specific antibodies to identify a series of molecules (fasciclins) required for axon guidance and fascicle formation along specific neural pathways, originally in grasshopper. The segmental nature of innervation makes fasciclin expression patterns in embryos ladder-like, with a repeating and bilateral pattern. Antifasciclin Fabs disrupt normal formation of the central nervous system in a manner similar to N-CAM defects in mice (Tomasiewicz et al. 1993; Cremer et al. 1994). Fasciclins are generally present in several stages of development, and participate in axon guidance, commissure formation, and specificity of synapse formation. *Drosophila* homologs have been identified and cloned. Among these, fasciclins II and III are members of the Ig superfamily. Fasciclin II (fasII) resembles vertebrate N-CAM in the presence of five Ig domains followed by two fibronectin type III domains, a transmembrane domain, and a short cytoplasmic tail. The sequences of the Ig domains in fasII and N-CAM are similar (Bieber et al. 1989). Fasciclin II is expressed on neurons in several pathways. Normal fasciculation and growth of these pathways is disrupted in *fasII* mutants (Lin et al. 1994). Similar molecules may also mediate plasticity of the nervous system in learning. ApCAM, molluskan homolog of FasII, is downregulated by decreased transcription and by specific endocytosis of existing molecules during long-term potentiation of the gill withdrawal response. The ApCAM functions as an apparent cell adhesion molecule, and mediates formation of axon bundles in neuron cultures. It is possible that a decrease in cell surface concentration of ApCAM is a prelude to increased axon growth and formation of the new synapses that marks the potentiation response (Mayford et al. 1992).

Fasciclin III (fasIII; Fig. 2) contains three Ig-like domains, followed by a 12-residue Pro-Ser-rich spacer. There is a transmembrane sequence and a 118-residue cytoplasmic tail which includes potential sites for Ser and Tyr phosphorylation (Snow et al. 1989). It functions as a homophilic cell adhesion molecule when expressed on nonadhesive host cells. FasIII functions as a specific targeting signal molecule in formation of defined neuromuscular junctions, i.e., it is required for specific axons to form synapses with specific muscle fibers. Deletion of FasIII can be compensated by other mechanisms, but ectopic expression causes "promiscuous" innervation of inappropriate muscle fibers (Chiba et al. 1995).

Other *Drosophila* Ig superfamily members include Amalgam, neuroglian, and Drtk. Neuroglian is an apparent homolog of the neural CAM-L1, with six Ig domains (C2 type), four fibronectin type III domains, a transmembrane sequence, and a short cytoplasmic tail (Bieber et al. 1989). It is expressed in

CNS, salivary gland, hindgut, peripheral nerve roots, and fasciculating axons. Deletions of the *Neuroglian* locus are embryonic lethal, but without obvious disruption of nervous system organization.

Amalgam is encoded by a gene in the *Antennapedia* complex, and includes a signal sequence and three Ig-like domains. The third domain is followed by a short C-terminal sequence (~15 residues), which resembles no known membrane attachment domain, so surface attachment may be through a receptor protein. Amalgam is expressed on the surface of various populations of preneural and neural cells during development, but deletions do not have an obvious phenotype (Seeger et al. 1988).

Dtrk encodes a type I membrane glycoprotein with six Ig domains and an intracellular tyrosine kinase domain (Pulido et al. 1992). The Dtrk protein mediates homophilic aggregation of expressing cells, and adhesion is accompanied by tyrosine phosphorylation of the protein, presumably by autophosphorylation. The protein shows extensive but highly patterned expression throughout development.

Other Fasciclins. Fasciclin I (fasI) is a GPI-anchored cell adhesion protein with four homologous 150-residue extracellular domains (Fig. 2). The sequence defines a gene family that now includes algal and mammalian homologs (Huber and Sumper 1994). FasI is expressed in differently spliced forms at various times in development. The differential splicing expresses 6- and 9- BP microexons and is postulated to modify adhesion properties (McAllister et al. 1992). FasI acts as a homophilic cell adhesion molecule, and in aggregation mixtures of cells expressing FasI and FasIII, the two types of cells sort out from each other. FasI is expressed by all pioneering neurons, and inactivation of FasI or its removal from cell surfaces by treatment with phospholipase C disrupts stereotypical migration of the pioneer cell growth cones (Jay and Keshishian 1990; Boyan et al. 1995). Fasciclin IV, now called semaphorin, is a family of proteins in *Drosophila*, and several vertebrate homologs are known. Semaphorins can function as adhesion modulators or inhibitors; collapsin, a chicken semaphorin, induces growth cone collapse in neurites. Semaphorins have a 600-residue extracellular domain with a Cys-rich N-terminal region and some forms have a single IgC2-like sequence (Kolodkin et al. 1993; Luo et al. 1993). There is a transmembrane sequence and a short cytoplasmic tail.

Leucine-Rich Repeat Proteins. The leucine-rich-repeat (LRR) is a 24-residue sequence that appears in multiple tandem arrays in a family of *Drosophila* surface molecules. The consensus includes Leu residues at positions 8, 11, 14, 16, 21, and 24. The known members of the family in *Drosophila* have 9 to 40 LRR repeats. The tandem repeats are flanked by Cys-rich regions with similarity to domains in the mammalian proteoglycan LRR proteins decorin and biglycan. Toll protein (Fig. 2), required for establishment of the dorsoventral axis, has 40 LRR repeats, a transmembrane domain, and an intracellular domain characteristic of vertebrate IL-1 receptors. It can function as a heterophilic cell adhesion protein (Keith and Gay 1990). Mutation of a Cys residue in one of the LRR-flanking sequences leads to constitutive activation of Toll, and relocalization of Dorsal protein from the nucleus (Kubota et al. 1993). Thus, Toll appears to mediate contact-dependent alteration of cellular localization of a differentiation factor. The mutation leads to altered glycosylation, and activation may proceed through aggregation of the Toll protein monomers (Kubota et al. 1995).

The LRR protein Connectin is regulated by *Ultrabithorax*, is GPI-anchored, and can function as a homophilic cell adhesion gene (Gould and White 1992; Nose et al. 1992). It is expressed on a distinct set of muscle cells and the growth cones of nerve fibers that innervate them. However, connectin is antiadhesive when ectopically expressed, showing a complex interaction with other cell adhesion systems (Nose et al. 1994). Toll is expressed at the same time, but on a different subset of muscle cells, and not on neurons. Thus, there is specificity in the expression patterns of these proteins.

Chaoptin is an LRR repeat cell adhesion molecule required for organization of microvilli in the rhabdomeres of the eye (Krantz and Zipursky 1990). Tartan is a transmembrane LRR protein with a novel intracellular sequence. It is expressed widely in developing embryos, in both anterior-posterior and dorsoventral stripes (hence its name). Homozygous loss of function mutations cause disorganization of the peripheral nervous system and muscles (Chang et al. 1993).

Cadherins. The cadherins are a family of homophilic cell adhesion proteins found primarily in adherens junctions and related structures (Kemler 1992). They are structurally related to the Ig superfamily, but have no apparent sequence similarity (Shapiro et al. 1995a). The relative orientation of the tandem cadherin Ig-like domains is regulated by Ca^{2+} ions at the domain interfaces. Cell-cell adhesion may be mediated by interaction of multiple molecules in a "zipper-like" arrangement (Shapiro et al. 1995a). Cadherins associate with microfilaments through cytoplasmic proteins called catenins. Cadherins are regulated in an "inside-out" manner; they are only active as cell adhesion molecules when associated with catenins and cytoskeletal elements. The adhesive complex can be inactivated by expression of the *v-src* kinase, accompanied by Tyr phosphorylation of catenins. In *Drosophila,* homologs of cadherins and catenins localize to junctional structures (Oda et al. 1993; Peifer et al. 1993). The function of the pattern formation gene *wingless* requires *armadillo* (a β-catenin), implying that organized histogenesis includes proper activation of the junctional complexes. A fly cadherin gene has been cloned and sequenced (Oda et al. 1994). Other cadherin-like proteins in flies include the products of *l(2)gl* and *fat*, a protein with 34 cadherin domains (Schmidt 1989; Mahoney et al. 1991; Grunwald 1993). Both of these genes function as tumor suppressors and are required for proper organization of the embryo. Neither of these proteins appears to be the junctional cadherin, and they may be homologs of related proteins involved in cell-substrate interactions (Grunwald 1993).

Integrins. Integrins are a diverse family of cell surface receptors that participate in cell-substrate and cell-cell interactions. These proteins consist of two subunits, each with a transmembrane domain and a short cytoplasmic sequence (Fig. 2). Some α subunits are proteolytically processed, with the products remaining disulfide-bonded, and the processed forms carry a domain homologous to several extracellular matrix proteins. Most α subunits have three to four copies of a metal-binding motif that may regulate activity. β subunits have a highly conserved N-terminal region of about 400 residues, followed by about 500 nonconserved residues, and an extracelular Cys-rich region with 56 Cys residues. In mammals, eight β subunits and at least 16 α subunits are known, and they can associate in diverse α-β combinations. Each combination has a characteristic set of ligands, which include laminin, collagen, fibronectin, and members of the Ig superfamily (Sanchez-Madrid and Corbi 1992; Clark and Brugge 1995; Lawrence et al. 1995). Integrin-mediated adhesions are essential for many cellular functions and mediate alterations of gene expression. Cross-linking of integrins through the extracellular domains modifies interactions of the β subunits with a complex of signaling proteins in a focal adhesion complex. These proteins include cytoskeletal-associated proteins such as α-actinin, vinculin, and talin, the tyrosine kinase focal adhesion kinase, and signal molecules such as Src, SOS, tensin, paxillin, Crk, and others. Presumably, ligand-induced clustering of the integrins modulates activity of these components to trigger cellular responses (Clark and Brugge 1995).

In *Drosophila*, an integrin family called PS is composed of a β subunit encoded by *myospheroid* and α subunits α_{PS1} and α_{PS2}. The latter is encoded by *inflated* and has alternatively spliced forms expressed at different times and locations in development. PS integrins can bind vitronectin and fibronectin, and homozygous loss-of-function mutations in the PS integrins result in embryonic lethality, with failures of muscle attachment, nervous system development, and dorsal closure (Brown et al. 1993). In metamorphosis, mutations cause deadhesion of pigment cells in the eye and failure of wing epithelia to bond, resulting in blistered wings. The availability of *Drosophila* models for integrin function has allowed the beginnings of functional dissection of the subunits, as well as determination of the developmental programs for each form (Zusman et al. 1993). A recent mutagenesis analysis of the intracellular sequence of β_{PS} has determined that the cytoplasmic tail is essential for integrin function and has identified residues important for proper development (Grinblat et al. 1993). Other *Drosophila* integrin genes are known, but are currently less well characterized.

8.5 Deuterostomes

8.5.1 Fertilization in Sea Urchins

Among the deuterostomes, adhesion has been studied in sea urchin, especially for sperm-egg interaction. A sequential series of cell-cell recognition events leads to sperm binding to and penetration of the jelly coat and vitelline layer, followed by membrane binding and fusion. This subject has been recently wonderfully reviewed (Foltz and Lennarz 1993), so only highlights will be mentioned here. An egg-derived peptide interacts with a sperm receptor to stimulate motility, an event analogous to the action of yeast sex pheromones, which prime the mating reaction. Contact of the sperm with the egg jelly coat stimulates the acrosome reaction in the sperm. Neither component in this initial cell adhesion is known for certain, but an N-glycosylated egg protein and a 210 -kDa sperm protein have been implicated. A 63-kDa GPI-anchored sperm protein with EGF repeats has also been implicated in the interaction. In sea urchin, the mechanism of penetration of the vitelline layer is not known, but in abalone (a mollusk) species-specific lysins dissolve the vitelline layer around the individual sperm (Shaw et al. 1995). Interaction of sea urchin sperm and egg membranes is through a heterologous species-specific interaction of sperm bindin and egg receptor protein. Bindin from the urchin *Stongylocentrotus purpuratus* has no obvious membrane attachment sequence, and may be a peripheral membrane protein. *Arbacia punctulata* bindin has a short hydrophobic sequence that may act as a fusogenic peptide, akin to viral fusion peptides, algal CAMs, and slime mold gp24 (Glabe and Clark 1991). The egg receptor from *S. purpuratus* has been cloned and sequenced. It consists of an extracellular domain that is Cys-rich and related to the Hsp70 class of chaperones. A Pro-rich sequence is proximal to the membrane. The 131-kDa polypeptide is highly N- and O-glycosylated and sulfated to give an apparent size of 350 kDa for the mature protein (Ohlendieck et al. 1993). Proteolytic fragments bind to sperm, but in a species-unspecific manner; whereas binding of the intact extracellular domain to sperm is species-specific. These results imply dual recognition: a nonspecific lectin-sugar interaction modified by specificity in the peptide sequence. Such a mode of binding would be compatible with other examples of carbohydrate-mediated modulation of binding in cell adhesion in yeast, sponge, slime molds, and vertebrate N-CAM (Rutishauser et al. 1988). The egg receptor protein presumably delivers a signal on sperm binding, resulting in an increase in intracellular Ca^{2+} and the cortical reaction. Although antibodies to the receptor can sometimes trigger these reactions, binding of purified bindin does not. This paradox remains to be resolved. As expected from their position on the evolutionary tree (Fig. 1), sea urchins have genes encoding proteins similar to other cell adhesion proteins, but their function has not been genetically or biochemically characterized (Mendoza et al. 1993).

9 Conclusions

Many of the nonvertebrate cell adhesion systems are recapitulated in development of vertebrates, for example, the presence of Ig domains and cdc10/SW14 sequences in yeast and vertebrate CAMS, and of FasI domains in algae and *Drosophila*. It therefore appears that development in multicellular and complex organisms uses cell adhesion systems that were present in a variety of simpler ancestors. It seems likely that precursors of most vertebrate cell adhesion systems existed in common ancestors among the protists, but that we have uncovered only a subset of them. The existence of multiple adhesion systems in *Dictyostelium* shows that cell-cell adhesion is not simple, even in relatively simple organisms. In yeast and *Chlamydomonas*, the sexual agglutinins are only one adhesion system of several whose existence may be inferred. Precursors of other types of cell adhesion systems may turn up in these organisms as gene-sequencing and adhesion studies proceed. It seems less likely that the plethora of cell adhesion systems in the higher animals could result from acquisition of the needed systems by horizontal gene transfer from simpler organisms or by convergent evolution (Sprague 1991; Bork and Doolittle 1992).

Thus, analysis of nonvertebrate adhesion systems is analogous to the fable of the blind men looking at the elephant: the characteristics of cell adhesion depend on the part of the elephant one examines. Nevertheless, we can start to draw a picture of the entire organism from the catalog of parts.

10 Characteristics of Cell Adhesion Proteins

The preceding exhausting (but not exhaustive) list of cell adhesion phenomena and proteins shows common features that reflect conserved origin and/or function. The proteins are modular, and share structural similarities including an extracellular amino termimus, and one or no transmembrane sequence. Positive feedback loops are a recurring finding where there is regulation of the expression of the adhesion proteins themselves.

10.1 Modules

Cell adhesion proteins are found primarily as monomeric proteins with large extracellular regions composed of repeats of a few types of structural domains. The functional domains start near the N-terminus of the mature protein. Some types of domains are recognizable from extremely simple sequence motifs, and can be traced through each of the kingdoms. There are two easily recognizable common themes in the presence of Cys-rich domains and Pro-rich domains. The function of the former is not clear, although the determination of the "sushi" structure and its appearance as the major feature of *Polysphondylium* gp70 implies a direct role in protein-protein interaction (Saito et al. 1994).

Although the spacing of Cys residues varies tremendously among the cell adhesion gene families, the ubiquity of the structures implies an essential function, and indeed such structures are also common in extracellular matrix proteins. In contrast, Pro-rich regions, and related glycosylated Pro-Ser-Thr-rich sequences usually appear close to surface attachment sites, and probably constitute stalks to extend the binding regions away from the cell (Noegel et al. 1986; Jentoft 1990; Ohlendieck et al. 1993; Chen 1995). An alternative use of Pro is to promote protein-protein interactions in pseudocrystalline arrays, as in animal collagens and extensins of plant cell walls and the algal cell adhesion proteins (Adair and Snell 1990). Both Pro-rich features result in extension of the length of the protein and resistance to proteases.

Figures 2 and 3 show cell adhesion molecules with several different types of extracellular domains. These have been identified by automated sequence searches, implying good conservation of key sequence motifs. Binding sites have been characterized for only a few of these domains (de Nobel et al., 1996; Siu and Kamboj 1990). The rate of evolution of cell adhesion sequences appears to be sufficiently slow that gene superfamilies can be easily recognized within the animal kingdom, and may be found with difficulty in other kingdoms founded in the Precambrian radiation. We may learn that more of these domains are related to each other, because three-dimensional structures are more strongly conserved than sequences. Among the common repeated domains, Ig, FnIII, and cadherin domains share no sequence similarity, but have three-dimensional structures sufficiently similar to imply a common ancestry (Doolittle 1995; but see Shapiro et al. 1995b for an opposing view). The elucidation of relationships between protist and animal cell adhesion domains therefore awaits determination of more three-dimensional structures.

Ig domains are broadly adaptable in their ability to bind diverse ligands, with binding regions composed of ends of the β barrels of several domains. Because the ends of the barrels are composed of peptide loops, and the compact domain structures are not altered by sequence changes in the loops, the number of potential binding site geometries and specificities is very large. Thus, there are homophilic and heterophilic adhesions, and ligands of Ig superfamily members, including the immunoglobulins themselves, appear to be unlimited in number or structure. In contrast, cadherins are homophilic only, and association is due to cohesion of antiparallel strands (Shapiro et al. 1995a, see below). It is provocative that the structure of four known adhesive domains, Ig, cadherin, fibronectin type III, and EGF are predominantly β-sheet structures. Again, three-dimensional structures of more cell adhesion proteins will be needed before we know if this is a general observation.

10.2 Cell Surface Association

Only a few modes of cell surface attachment are found within the cell adhesion molecules we have described. Many are type I transmembrane proteins, with the amino terminal extracellular and the carboxy terminal cytoplasmic. No example of type II (C extracellular, N intracellular) or multipass proteins is cited as a cell adhesion protein. The proteins are mostly monomeric, but multimeric association through disulfides or ligand induced cross-linking is also seen. In the type I proteins, the cytoplasmic sequences interact with the cytoskeleton and/or signal transducing machinery, although the pathways are often not yet clear.

A second mode of attachment is through GPI anchors. This structure is present on proteins for which no signal transduction event is known for any GPI-anchored cell adhesion protein, although they can facilitate signaling by localization of signals, as in yeast and synapses (Lipke and Kurjan 1992). In the case of *Dictyostelium* gp80, the GPI anchorage stabilizes the protein by reducing turnover.

Several cell adhesion proteins have no mode of cell surface attachment that can be inferred from the sequence. In some cases (sponge AF-binding factors), the interaction is through association with cell surface proteins or lipids. In other cases, there is disulfide bonding to cell surface proteins (yeast a-agglutinin, integrin α subunits). Finally, some proteins have unknown mechanisms of membrane association, but may have short hydrophobic sequences akin to viral fusogenic peptides (Loomis and Fuller 1990; Glabe and Clark1991; McAllister et al. 1992; Borst et al. 1995).

10.3 Role of Ca^{2+}

Ca-dependent cell adhesion is seen in the cadherins, EGF domains, the sponge AFs, and in yeast flocculation. In the former three systems, the ion maintains the relative orientation of the adhesive domains to promote multiple, regular interactions (Misevic and Burger 1993; Rao et al. 1995; Shapiro et al. 1995a). Such arrays are indeed crystalline for cadherin and EGF structures, and could be so for the sponge factor. The latter is unique in that a carbohydrate-carbohydrate association is promoted. In all three cases, multiple contacts greatly increase the strength of cell-cell interactions through inherently weak bonds. Indeed, the modularity of these proteins stems from the need for multiplicity and regularity of interactions. In other cases, the mode of Ca^{2+} activation is unknown (Diehl-Seifert et al. 1985b; Robitzki et al. 1990; Fontana 1993).

10.4 Binding Characteristics

Where binding affinities have been measured, they are in the range of 10^7 L^{-1} to 10^9 L^{-1}. In this range, binding is reversible by dilution in many cases. In cases of multiple interactions, association constants are higher (10^{10}–10^{14} L^{-1}), and associations are essentially irreversible. Site numbers for monomeric cell adhesion proteins are in the range of 10^4 to 10^5 per cell, but are probably much higher for the junctional proteins such as cadherins and integrins.

11 Extracellular Matrix Interactions

I have neglected many extracellular matrix interactions. The panoply of protein-protein, protein-carbohydrate, and carbohydrate-carbohydrate interactions that make up metazoan ECM have been often reviewed (Scott 1992; Engel et al. 1994; Murray et al. 1995), and may be relevant to the cell adhesion systems described.

12 Role of Lectins and Carbohydrates

A great mystery remains unsolved. There is as yet no unifying explanation for the role of carbohydrate-binding proteins in cell adhesion. Although sugars are known to be important for function of cell adhesion systems in *Plasmodium*, slime molds, yeast, sponge, and sea urchins, a common mechanism is not yet obvious. The discovery of selectins as transient immunological specifiers of tissue location and inflammation in the vertebrates has as yet no counterpart among the nonvertebrates. One possibility is that the lectin interactions involve elements of the extracellular matrix, as in discoidin and the sponge lectin (Feit 1994; Diehl-Seifert et al. 1985b). Alternatively, the model of N-CAM, in which there is a lectin-like region, and sialylation modifies the adhesive properties, may prove more general (Rutishauser et al. 1988). Sialylation is important in sponge aggregation, sperm-egg interaction, and also appears to modify host-pathogen interactions in *Plasmodium*. There is now an example of a mammalian cell adhesion protein, CD2, in which N-linked carbohydrate maintains structure of the binding site (Wyss et al. 1995). As originally pointed out by Roseman, the possible specific polysaccharide structures are practically infinite, and provide great opportunity for specificity and regulation of activity (Roseman 1970).

13 Signal Transduction and Cytoplasmic Domains

Cytoplasmic domains of cell adhesion proteins transduce signals. The transduction may appear to be a secondary role (as in contact sites A of *Dictyostelium*) or the primary function, with cell association secondary (Notch family). This difference is probably more in our perception than in critical functions of the molecule: it all depends on which molecular function, adhesion or signaling, was assayed first. In the *Chlamydomonas* agglutination system and in slime mold, cAMP is implicated as an induced or complementary messenger. In other cell adhesion systems direct protein-protein interactions mediate the signals. Examples include the yeast pheromone system (which is only casually related to adhesion), *Notch/Delta*- and *Toll*-dependent systems, in which a transcription factor appears to be released from membrane association and translocated to the nucleus, and Ig superfamily members with SH3 and protein kinase domains. In the cases of the cadherins, integrins, and the Crumbs protein, the cell adhesion proteins serve as organization points for the cytoskeleton, and help maintain cell polarity. Adhesive interactions can thus modify the asymmetry of the cell and orientation within the tissue.

The final picture is of a diverse set of proteins and glycoproteins, which mediate adhesion through specific protein-protein interactions. Interaction with carbohydrates may modify the binding in some cases. Cell-cell adhesion molecules are made up of a few dozen types of functional domains, and a larger set of domains appears to participate in cell-substrate and ECM-ECM interactions. In the case of the GPI-anchored proteins, no intracellular ramifications are seen. In systems incorporating transmembrane proteins, extracellular binding stimulates many different kinds of intracellular signals.

Acknowledgments. I thank Anthony Walker for help with the research and Rebecca Crespo for outstanding typing. I thank Katherine Lyser for comments on the manuscript, and Mitch Sogin for helpful discussions.

References

Adair WS, Snell WJ (1990) The *Chlamydomonas reinhardtii* cell wall: structure, strutcure, biochemistry, and molecular biology. In: Organization and assembly of plant and animal extracellular matrix. Academic Press, New York, pp 14–85

Adams JH, Sim BKL, Dolan SA, Fan X, Kaslow DC, Miller LH (1994) A family of erythrocyte binding proteins of malaria parasites. Proc Natl Acad Sci 89: 7085–7089

Aiba K, Yanagisawa K, Urushihara H (1993) Distribution of gp138, a cell surface protein responsible for sexual cell fusion, among cellular slime moulds. J Gen Microbiol 139: 279–285

Artavanis-Tsakonas S, Matsuno K, Fortini ME (1995) Notch signalling. Science 268: 225–232

Baffi RA, Shenbagamurthi P, Terrance K, Becker JM, Naider F, Lipke PN (1984) Different structure-function relationships for α-factor-induced morphogenesis and agglutination in *Saccharomyces cerevisiae*. J Bacteriol 158: 1152–1156

Baffi RA, Becker JM, Lipke PN, Naider F (1985) Structure-activity relationship in the dodecapeptide α-factor of *Saccharomyces cerevisiea*: position 6 analogues are poor inducers of agglutinability. Biochemistry 24: 3332–3337

Barnes RD (1980) Invertebrate zoology. Saunders, Philadelphia

Barth A, Muller-Taubenberger A, Taranto P, Gerisch G (1994) Replacement of the phospholipid-anchor in the contact site Aglycoprotein of *D. discoideum* by a transmembrane region does not impede cell adhesion but reduces residence time on the cell surface. J Cell Biol 124: 205–215

Baruch DI, Pasloske BL, Singh HB, Bi X, Ma XC, Feldman M, Taraschi TF, Howard RJ (1995) Cloning of the *P. falciparum* gene encoding PfEMP1, a malarial variant antigen and adherence receptor on the surface of parasitized human erythrocytes. Cell 82: 77–87

Bass SH, Mulkerrin MG, Wells JA (1991) A systematic mutational analysis of hormone binding determinants in the human growth hormone receptor. Proc Natl Acad Sci USA 88: 4498–4502

Bendel CM, Hostetter MK (1993) Distinct mechanisms of epithelial adhesion for *Candida albicans* and *Candida tropicalis*. J Clin Invest 92: 1840–1849

Bieber AJ, Snow PN, Hortsch M, Patel NH, Jacobs JR, Traquina ZR, Schilling J, Goodman C (1989) *Drosophila* neuroglian: a member of the immunoglobulin superfamily with extensive homology to the vertebrate neural adhesion molecule L1. Cell 59: 447–460

Bonner JT (1982) Comparative biology of the cellular slime molds. In: Loomis WF (ed) The development of *Dictyostelium discoideum*. Academic Press, New York, pp 1–34

Bork P, Doolittle RF (1992) Proposed acquisition of an animal protein domain by bacteria. Proc Natl Acad Sci USA 89: 8990–8994

Borst P, Bitter W, McCullough R, Van Leeuwen F, Rudenko G (1995) Antigenic variation in malaria. Cell 82: 1–4

Boyan G, Therianos JL, Williams JL, Reichert H (1995) Axonogenesis in the embryonic brain of the grasshopper *Schistoceria gregaria*: an identified cell analysis of early brain development. Development 121: 75–86

Brar SK, Siu C-H (1993) Characterization of the cell adhesion molecule gp24 in *Dictyostelium discoideum*. Mediation of cell-cell adhesion via a Ca(2+)-dependent mechanism. J Biol Chem 268: 24902–24909

Breuer W, Siu C-H (1981) Identification of endogenous binding proteins for the lectin discoidin-I in *Dictyostelium discoideum*. Proc Natl Acad Sci USA 78: 2115–2119

Brown NH, Bloor JW, Dunin-Borkowski A, Martin-Bermudo MD (1993) Integrins and morphogenesis. Development (Suppl): 177–183

Calderone RA (1993) Molecular interactions at the interface of *Candida albicans* and host cells. Arch Med Res 24: 275–279

Cappellaro C, Hauser K, Mrsa V, Watzele M, Watzele G, Gruber C, Tanner W (1991) *Saccharomyces cerevisiae* **a**- and α-agglutinin: characterization of their molecular interaction. EMBO J 10: 4081–4088

Cappellaro C, Baldermann C, Rachel R, Tanner W (1994) Mating type-specific cell-cell recognition of *Saccharomyces cerevisiae*: cell wall attachment and active sites of **a**- and α-agglutinin. EMBO J 13: 4737–4744

Carroll SB (1995) Homeotic genes and the evolution of the arthropods and chordates. Nature 376: 479–485

Cauldwell CB, Henkart P, Humphreys T (1973) Physical properties of sponge aggregation factor. A unique proteoglycan complex. Biochemistry 12: 3051–3055

Chang Z, Price BD, Bockheim S, Boedigheimer MJ, Smith R, Laughon A (1993) Molecular and genetic characterization of *Drosophila tartan* gene. Dev Biol 160: 315–332

Chen M-H, Shen Z-M, Bobin S, Kahn PC, Lipke PN (1995) Structure of *Saccharomyces cerevisiae* α-agglutinin: evidence for a yeast cell wall protein with multiple immunoglobulin-like domains with atypical disulfides. J Biol Chem 270: 26168–26177

Chiba A, Snow P, Keshishian H, Hotta Y (1995) Fasciclin III as a synaptic target recognition molecule in *Drosophila*. Nature 374: 166–168

Chitnis CE, Miller LH (1994) Identification of the erythrocyte binding domains of the *Plasmodium vivax* and *Plasmodium knowlesi* proteins involved in erythrocyte invasion. J Exp Med 180: 497–506

Choi AH, Siu C-H (1987) Filopodia are enriched in a cell cohesion molecule of Mr 80,000 and participate in cell-cell contact formation in *Dictyostelium discoideum*. J Cell Biol 104: 1375–1387

Chothia C (1994) Protein families in the metazoan genome. Development (Suppl): 27–33

Chuenkova M, Pereira ME (1995) *Trypanosoma cruzi trans*-sialidase: enhancement of virulence in a murine model of Chaga's disease. J Exp Med 181: 1693–1703

Clark EA, Brugge JS (1995) Integrins and signal transduction: the road taken. Science 268: 233–239

Conway MS (1994) Why molecular biology needs paleontology. Development (Suppl): 1–13

Corbin V, Michelson AM, Abmayr SM, Neel V, Alcamo E, Maniatis T, Young MW (1991) A role for the *Drosophila* neurogenic genes in mesoderm differentiation. Cell 67: 311–323

Cremer H, Lange R, Cristoph A, Plomann M, Vopper G, Roes J, Brown R, Baldwin S, Kraemer, Scheff S, Barthels D, Rajewwski K, Willie W (1994) Inactivation of the N-CAM gene in mice results in size reduction of the olfactory bulb and deficits in spatial learning. Nature 367: 455--459

Cross GA, Takle GB (1993) The surface *trans*-sialidase family of *Trypnosoma cruzi*. Annu Rev Microbiol 47: 385–411

De Nobel H, Lipke PN (1994) Is there a role for GPI's in yeast cell wall assembly? Trends Cell Biol 4: 42–45

De Nobel H, Pike J, Lipke PN, Kurjan J (1995) Genetic analysis of **a**-agglutinin in *S. cerevisiae* and identification of a plasmid expression mutant. Mol Gen Genet 247: 409–415

De Nobel H, Lipke PN, Kurjan J (1996) Identification of a ligand binding site in an immunoglobulin fold domain of the *Saccharomyces cerevisiae* α-agglutinin. Cell Mol Biol 7: 143–153

Desbarats L, Brar SK, Siu C-H (1994) Involvement of cell-cell adhesion in the expression of the cell cohesion molecule gp80 in *Dictyostelium discoideum*. J Cell Sci 107: 1705–1712

Diehl-Seifert B, Kurelec B, Zahn RK, Dorn A, Jericevic B, Uhlenbruck G, Müller WE (1985a) Attachment of sponge cells to collagen substrata: effects of a collagen assembly factor. J Cell Sci 79: 271–285

Diehl-Seifert B, Uhlenbruck G, Geisert M, Zahn K, Müller WE (1985b) Physiochemical and functional characterization of the polymerization process of the *Geodia cydonium* lectin. Eur J Biochem 147: 517–523

Doolittle RF (1995) The multiplicity of domains in proteins. Annu Rev Biochem 64: 287–314

Engel J, Efimov VP, Maurer P (1994) Domain organization of extracellular matrix proteins and their evolution. Development (Suppl): 35–42

Fang H, Higa M, Suzuki K, Aiba K, Urushihara H, Yanagisawa K (1993) Molecular cloning and characterization of two genes encloding gp138, a cell surface glycoprotein involved in the sexual cell fusion of *Dictyostelium discoideum*. Dev Biol 156: 201–208

Feit IN (1994) Cell prints on the surface of the slug of *Dictyostelium discoideum*: a Nessler-positive matrix substance. Dev Biol 164: 345–360

Foltz KR, Lennarz WJ (1993) The molecular basis of sea urchin gamete interactions in the egg plasma membrane. Dev Biol 158: 46–61

Fontana DR (1993) Two distinct adhesion systems are responsible for EDTA-sensitive adhesion in *Dictyostelium discoideum*. Differentiation 53: 139–147

Gamulin V, Rinkevich B, Schäcke H, Kruse M, Müller IM, Müller EG (1994) Cell adhesion receptors and nuclear receptors are highly conserved from the lowest metazoa (marine sponges) to vertebrates. Hoppe-Seyler's Z Biol Chem 375: 583–588

Gao EN, Shier P, Siu C-H (1992) Purification and partial characterization of a cell adhesion molecule (gp150) involved in postaggregation stage cell-cell binding in *Dictyostelium discoideum*. J Biol Chem 267: 9409–9415

Gerisch G, Weinhart U, Bertholdt G, Claviez M, Stadler J (1985) Incomplete contact site A glycoprotein in HL220, A modB mutant of *Dictyostelium discoideum*. J Cell Sci 73: 49–68

Gettner SN, Kenyon C, Reichardt LF (1995) Characterization of β pat-3 heterodimers, a family of essential integrin receptors in *C. elegans*. J Cell Biol 129: 1127–1141

Ghysen A, Dambly-Chaudiere C, Jan LY, Jan Y-N (1993) Cell interactions and gene interactions in peripheral neurogenesis. Genes Dev 7: 723–733

Glabe CG, Clark D (1991) The sequence of *Arbacia punctulata* binding cDNA and implications for the structural basis of species-specific sperm adhesion and fertilization. Dev Biol 143: 282–288

Glee PM, Sundstrom P, Hazen KC (1995) Expression of surface hydrophobic proteins by *Candida albicans* in vivo. Infect Immun 63: 1373–1379

Goodenough U (1991) *Chlamydomonas* mating interactions. In: Dworkin M (ed) Microbial cell interactions. Am Soc Microbiol, Washington, pp 71–112

Gould AP, White RAH (1992) *Connectin*, a target of homeotic gene control in *Drosophila*. Dev 116: 1163–1174

Gramzow M, Bachmann M, Uhlenbruck G, Dorn A, Müller WE (1986) Identification and further characterization of the specific cell binding fragment from sponge aggregation factor. J Cell Biol 102: 1344–1349

Greenwald I (1989) Cell-cell interactions that specify cell fate in *C. elegans* development. Trends Genet 5: 237–241

Grinblat Y, Zusman S, Yee G, Hynes RO, Kafatos FC (1993) Functions of the cytoplasmic domain of the b_{PS} integrin subunit during *Drosophila* development. Development 120: 91–102

Grunwald GB (1993) The structural and functional analysis of cadherin calcium-dependent cell adhesion molecules. Curr Opin Cell Biol 5: 797–805

Harloff C, Gerisch G, Noegel AA (1989) Selective elimination of contact site A protein of *Dictyostelium discoideum* by gene disruption. Genes Dev 3: 2011–2019

Henderson ST, Gao D, Lambie EJ, Kimble J (1994) *Lag-2* may encode a signalling ligand for the GLP-1 and LIN-12 receptors of *C. elegans*. Development 120: 2913–2924

Herrera EM, Ming M, Ortega-Barria E, Pereira ME (1994) Mediation of *Trypnosoma cruzi* by heparan sulfate receptors on host cells and penetrin counter-receptors on the trypanosomes. Mol Biochem Parasitol 65: 73–83

Hill RJ, Sternberg PW (1993) Cell fate patterning in *C. elegans* vulval development. Development (Suppl): 9–18

Hinkle G, Leipe DD, Nerad TA, Sogin ML (1994) The unusually long small subunit ribosomal RNA of *Phreatamoeba balamuthi*. Nucl Acids Res 22: 465–469

Hirabayashi J, Satoh M, Kasai K (1992) Evidence that Caenorhabditis elegans 32-kDa beta-galactoside-binding protein is homologous to vertebrate beta-galactoside-binding lectins. cDNA cloning and deduced amino acid sequence. J Biol Chem 267: 15485–15490

Hirano T, Yamada H, Miyazaki T (1985) Direct implication of surface mannosyl residues in cell adhesion of *Dictyostelium discoideum*. J Biochem (Tokyo) 98: 199–208

Holtfreter J (1943) Gewebeaffinität, ein Mittel der embryonalen Formbildung. Arch Exp Zelif 23: 169–209. Translated In: Willier BH, Oppenheimer JM (eds) Foundations of experimental embryology. Prentice Hall, NY, 1982

Hoyer LL, Scherer S, Schatzman AR, Livi GP (1995) *Candida albicans ALS1*: domains related to a *Saccharomyces cerevisiae* sexual agglutinin separated by a motif. Mol Microbiol 15: 39–54

Huber O, Sumper M (1994) Algal-CAMS: isoforms of a cell adhesion molecule in embryos of the alga *Volvox* with homology to *Drosophila* fasciclin I. EMBO J 13: 4212–4222

Hutter H, Schnabel R (1994) *glp-1* and inductions establishing embryonic axes in *C. elegans*. Development 120: 2051–2064

Jackson CL, Hartwell LH (1990) Courtship in *S. cerevisiae*: both cell types choose mating partners by responding to the strongest pheromone signal. Cell 63: 1039–1051

Jackson CL, Konopka JB, Hartwell LH (1991) *S. cerevisiae* alpha pheromone receptors activate a novel signal transduction pathway for mating partner discrimination. Cell 67: 389–402

Jay DG, Keshishian H (1990) Laser inactivation of fascilin I disrupts axon adhesion of grasshopper pioneer neurons. Nature 348: 548–550

Jentoft N (1990) Why are proteins O-glycosylated? Trends Biochem Sci 15: 291–295

Jumblatt JE, Schlup V, Burger MM (1980) Cell-cell recognition: specific binding of Microciona sponge aggregation factor to homotypic cells and the role of calcium ions. Biochemistry 19: 1038–1042

Kamboj RK, Gariepy J, Siu C-H (1989) Identification of an octapeptide involved in homophilic interaction of the cell adhesion molecule gp80 of *Dictyostelium discoideum*. Cell 59: 615–625

Kaniga K, Tucker S, Trollinger D, Galan JE (1995) Homologs of *Shigella IpB* and *IpC* invasins are required for *Salmonella typhimurium* entry into epithelial cells. J Bacteriol 177: 3965–3971

Keith FJ, Gay NJ (1990) The *Drosophila* membrane protein Toll can function to promote cell adhesion. EMBO J 9: 4299–4306

Kemler R (1992) Classical cadherins. Sem Cell Biol 3: 149–155

Klis FM (1994) Review: cell wall assembly in yeast. Yeast 10: 851–869

Kolodkin AL, Mathis DJ, Goodman CS (1993) The *semaphorin* genes encode a family of transmembrane and secreted growth cone guidance molecules. Cell 75: 1389–1399

Krantz DE, Zipursky SL (1990) *Drosophila* chaoptin, a member of the leucine-rich repeat family is a photoreceptor cell-specific adhesion molecule. EMBO J 9: 1969–1977

Kubota K, Keith FJ, Gay NJ (1993) Relocalization of *Drosophila* dorsal protein can be induced by a rise in cytoplasmic calcium and the expression of constitutively active but not wild-type Toll receptors. Biochem J 296: 497–503

Kubota K, Keith FJ, Gay NJ (1995) Wild type and constitutively activated forms of *Drosophila* Toll receptor have different patterns of N-linked glycosylation. FEBS Lett 365: 83–86

Kuchler K, Dohlman HG, Thorner J (1993) The **a**-factor transporter (*STE6* gene product) and cell polarity in the yeast *Saccharomyces cerevisiae*. J Cell Biol 120: 1203–1215

Kuhns WJ, Bramson S, Simpson TL, Burkart W, Jumblatt J, Burger MM (1980) Fluorescent antibody localization of *Microciona prolifera* aggregation factor and its baseplate component. Eur J Cell Biol 23: 73–79

Kurjan J (1993) The pheromone response pathway in *Saccharomyces cerevisiae*. Annu Rev Genet 27: 147–179

Lawrence MB, Berg EL, Butcher EC, Springer TA (1995) Rolling of lymphocytes and neutrophils on peripheral node addressin and subsequent arrest on ICAM-1 in shear flow. Eur J Immunol 258: 1025–1031

Leung-Hagesteijn C, Spence AM, Stern BD, Zhou Y, Su MW, Hedgecock EM, Culot JG (1992) UNC-5, a transmembrane protein with immunoglobulin and thrombospondin type 1 domains, guides cell and pioneer axon migrations in *C. elegans*. Cell 71: 289–299

Lieber T, Wesley CS, Alcamo E, Hassel B, Krane JF, Campos-Ortega JA, Young MW (1992) Single amino acid substitutions of notch and delta modify *Drosophila* development and affect cell adhesion in vitro. Neuron 9: 847–859

Lin DM, Fetter RD, Kopczynski C, Grenningloh G, Goodman CS (1994) Genetic analysis of fasiculin II in *Drosophila*: defasciculation, refasciculation, and altered fasciculation. Neuron 13: 1055–1069

Lipke PN, Kurjan J (1992) Sexual agglutinins in budding yeasts: structure, function and regulation of yeast cell adhesion proteins. Microbiol Rev 56: 180–194

Lipke PN, Terrance K, Wu Y-S (1987) Interaction of α-agglutinin with *Saccharomyces cerevisiae* **a** cells. J Bacteriol 169: 483–488

Lipke PN, Chen M-H, De Nobel H, Kurjan J, Kahn PC (1995) Homology modeling of an immunoglobulin-like domain from the *Saccharomyces cerevisiae* adhesion protein α-agglutinin. Prot Sci. 4: 2168–2178

Loomis WF, Fuller DL (1990) A pair of tandemly repeated genes code for gp24, a putative adhesion protein of *Dictyostelium discoideum*. Proc Natl Acad Sci USA 87: 886–890

Lu CF, Montijn RC, Brown JL, Klis F, Kurjan J, Bussey H, Lipke PN (1995) Glycosyl phosphatidylinositol-dependent cross-linking of α-agglutinin and β1, 6-glucan in the *S. cerevisiae* cell wall. J Cell Biol 128: 333–340

Lundquist EA, Herman RK (1994) The mec-8 gene of *Caenorhabditis elegans* affects muscle and sensory neuron function and interacts with three other genes: unc-52, smu-1 and smu-2. Gene 138: 83–101

Luo Y, Raible D, Raper JA (1993) Collapsin: a protein in brain that induces the collapse and paralysis of neoronal growth cones. Cell 75: 217–227

Mahoney PA, Weber U, Onofrechuk P, Biessmann H, Bryant PJ, Goodman CS (1991) The *fat* tumor suppressor gene in *Drosophila* encodes a novel member of the cadherin superfamily. Cell 67: 853–868

Manabe R, Saito T, Kumazaki T, Sakaitani T, Nakata N, Ochiai H (1994) Molecular cloning and the COOH-terminal processing of Gp64, a putative cell-cell adhesion protein of the cellular slime mold *Polysphondylium pallidum*. J Biol Chem 269: 528–535

Marcantonio EE, Hynes RO (1988) Antibodies for the conserved cytoplasmic domain of the integrin b_1 subunit react with proteins in vertebrates, invertebrates, and fungi. J Cell Biol 106: 1765–1772

Margoliash E, Schenck JR, Hargie MP, Burokas S, Richter WR, Barlow GH, Moscona AA (1965) Characterization of specific cell aggregating materials from sponge cells. Biochem Biophys Res Commun 20: 383–388

Mayford M, Barzilai A, Keller F, Schacher S, Kandel ER (1992) Modulation of an NCAM related adhesion molecule in long-term synaptic plasticity in *Aplysia*. Science 256: 638–644

McAllister L, Rehm EJ, Goodman CS, Zinn K (1992) Alternative splicing of microexons creates multiple forms of insect adhesion molecule fasciclin I. J Neurosci 12: 895–905

McCourtie M, Douglas LJ (1985) Extracellular polymer of *Candida albicans*: isolation, analysis, and role in adhesion. J Gen Microbiol 131: 495–503

Mendoza LM, Nishioka D, Vacquier VD (1993) A GPI-anchored sea urchin sperm membrane protein containing EGF domains is related to human uromodulin. J Cell Biol 121: 1291–1297

Misevic GN, Burger MM (1986) Reconstitution of high cell binding affinity of a marine sponge aggregation factor by cross-linking of small low affinity fragments into a large polyvalent polymer. J Biol Chem 261: 2853–2859

Misevic GN, Burger MM (1993) Carbohydrate-carbohydrate interations of a novel acidic glycan can mediate sponge cell adhesion. J Biol Chem 268: 4922–4929

Moscona AA (1957) The development in vitro of chimeric aggregates of dissociated embryonic chick and mouse cells. Proc Natl Acad Sci USA 43: 184–94

Müller WE, Müller I, Zahn RK, Kurelec B (1976) Species-specific aggregation factor in sponges. VI. aggregation factor receptors from the cell surface. J Cell Sci 21: 227-241

Müller WE, Arendes J, Kurelec B, Zahn RK, Müller I (1977) Species-specific aggregation factor in sponges. Sialyltransferase associated with aggregation factor. J Biol Chem 252: 3836–3842

Müller WEG, Zahn RK, Conrad J, Kurelec B, Uhlenbruck G (1982) Aggregation of sponge cells: stage dependent, distinct adhesion mechanisms in *Cliona celata*. Eur J Cell Biol 28: 243–250

Müller WE, Rottmann M, Diehl-Seifert B, Kurelec B, Uhlenbruck G, Schröder HC (1987) Role of the aggregation factor in the regulation of phosphoinositite metabolism in sponges. Possible consequences on calcium efflux and on mitogenesis. J Biol Chem 262: 9850–9858

Murray MA, Fessler LI, Palka J (1995) Changing distribution of extracellular matrix components during early wing morphogenesis in *Drosophila*. Dev Biol 168: 150–165

Noegel A, Harloff C, Hirth P, Merkl R, Modersitzki M, Stadler J, Weinhart U, Westphal M, Gerisch G (1985) Probing an adhesion mutant of *Dictyostelium discoideum* with cDNA clones and monoclonal antibodies indicates a specific defect in the contact site A glycoprotein. EMBO 4: 3805–3810

Noegel A, Gerisch G, Stadler J, Westphal M (1986) Complete sequence and transcript regulation of a cell adhesion protein from aggregating *Dictyostelium* cells. EMBO J 5: 1473–1476

Nose A, Mahajan VB, Goodman CS (1992) Connectin: a homophilic cell adhesion molecule expressed on a subset of muscles and the motoneurons that innervate them in *Drosophila*. Cell 70: 553–567

Nose A, Takeichi M, Goodman CS (1994) Ectopic expression of connectin reveals a repulsive function during growth cone guidance and synapse formation. Neuron 13: 525–539

Oda H, Uemura T, Shiomi K, Nagafuchi A, Tsukita S, Takeichi M (1993) Identification of a *Drosophila* homolog of a-catenin and its association with the *armadillo* protein. J Cell Bio 121: 1133–1140

Oda H, Uemura T, Harada Y, Iwai Y, Takeichi M (1994) A *Drosophila* homolog of cadherin associated with armidillo and essential for embryonic cell adhesion. Dev Biol 165: 716–726

Ohlendieck K, Dhume ST, Partin JS, Lennarz WS (1993) The sea urchin egg receptor for sperm: isolation and characterization of the intact, biological receptor. J Cell Biol 122: 887–895

Ollert MW, Sohnchen R, Korting HC, Ollert U, Brautigam S, Brautigam W (1993) Mechanisms of adherence of *Candida albicans* to cultured human epithelial cells. Infect Immun 61: 4560–4568

Ortega-Barria E, Pereira MEA (1991) A novel *T. Cruzi* heparin-binding protein promotes fibroblast adhesion and penetration of engineered bacteria and trypanosomes into mammalian cells. Cell 67: 411–421

Parish CR, Jakobsen KB, Coombe DR, Bacic A (1991) Isolation and characterization of cell adhesion molecules from the marine sponge, *Ophlitaspongia tenuis*. Biochim Biophys Acta 1073: 56–64

Patterson DJ, Sogin ML (1992) Eukaryotic origin and protistan diversity. In: Hartman H, Matsuno K (eds) The origins and evolution of the cell. World Scientific, Singapore, pp 13–46

Peifer M, Orsulic S, Pai L-M, Loureiro J (1993) A model system for cell adhesion and signal transduction in *Drosophila*. Development (Suppl): 163–176

Pendrak ML, Klotz SA (1995) Adherence of *Candida albicans* to host cells. FEMS Microbiol Lett 129: 103–113

Pruitt RE, Hulskamp M, Kopczak SD, Ploense S, Schneitz K (1993) Molecular genetics of cell interactions in *Arabidopsis*. Development (Suppl): 77–84

Pulido D, Campuzano S, Koda T, Modollel J, Barbacid M (1992) *Dtrk*, a *Drosophila* gene related to the *trk* family of neurotrophin receptors, encodes a novel class of neural cell adhesion molecule. EMBO J 11: 391–404

Rao Z, Handford P, Mayhew M, Knott V, Brownlee GG, Stuart D (1995) The structure of a Ca(2 +)-binding epidermal growth factor-like domian: its role in protein-protein interactions. Cell 82: 131–141

Rebay I, Fleming RJ, Fehon RG, Cherbas L, Cherbas P, Artavanis-Tsakonas S (1991) Specific EGF repeats of notch mediate interactions with selta and serrate: implications for notch as a multifunctional receptor. Cell 67: 687–699

Robitzki A, Schröder HC, Ugarkovic D, Gramzow M, Fritsche U, Batel R, Müller WE (1990) cDNA structure and expression of calpactin, a peptide involved in Ca2(+)-dependent cell aggregation in sponges. Biochemistry 271: 415–420

Rogalski TM, Williams BD, Mullen GP, Moerman DG (1993) Products of the *unc-52* gene in *Caenorhabditis elegans* are homologous to the core protein of the mammalian basement membrane heparan sulfate proteoglycan. Genes Dev 7: 1471–1484

Roseman S (1970) The synthesis of complex carbohydrates by multiglycosyl transferase systems and their potential role in intercellular adhesion. Chem Phys Lipids 5: 270–297

Rutishauser U, Acheson A, Hall AK, Mann DM, Sunshine J (1988) The neural cell adhesion molecule (NCAM) as a regulator of cell-cell interactions. Science 240: 53–57

Saito T, Ochiai H (1993) Evidence for a glycolipid anchor of gp64, a putative cell-cell adhesion protein of *Polysphondylium pallidum*. Eur J Biochem 218: 623–628

Saito T, Kumazaki T, Ochiai H (1993a) A purification method and *N*-glycosylation sites of a 36-cysteine-containing, putative cell-cell adhesion glycoprotein gp64 of the cellular slime mold, *Polysphondylium pallidum*. Eur J Biochem 211: 147–155

Saito T, Small L, Goodenough UW (1993b) Activation of adenylyl cyclase in *Chlamydomonas reinhardtii* by adhesion and heat. J Cell Biol 122: 137–147

Saito T, Kumazaki T, Ochiai H (1994) Assignment of disulfide bonds in gp64, a putative cell-cell adhesion protein of *Polysphondylium pallidum*. J Biol Chem 269: 28798–28802

Sanchez-Madrid F, Corbi AL (1992) Leukocyte integrins: structure, function, and regulation of activity. Sem Cell Biol 3: 199–210

Sano K, Tanihara H, Heimark RL, Obata S, Davidson M, St. John T, Taketani S, Suzuki S (1993) Protocadherins: a large family of cadherin-related molecules in central nervous system. EMBO J 12: 2249–2256

Schäcke H, Rinkevich B, Gamulin V, Müller IM, Müller WE (1994) Immunoglobulin-like domain is present in the extracellular part of the receptor tyrosine kinase from the marine sponge geodia cydonium. J Mol Recognit 7: 273–276

Schilling KM, Bowen WH (1992) Glucans synthesized in situ in experimental salivary pellicle function as specific binding sites for *Streptococcus mutans*. Infect Immun 60: 284–295

Schmidt O (1989) A recessive tumor gene function in *Drosophila* is involved in cell adhesion. J Neurogenet 5: 95–98

Schröder HC, Amberger V, Renneisen K, Bachmann M, Kurelec B, Uhlenbruck G, Müller WE (1989) Protein kinase C phosphorylates the sponge aggregation receptor after its binding to the homologous aggregation factor. Eur J Cell Biol 48: 142–149

Scott JE (1992) Supramolecular organization of extracellular matrix glycosaminioglycans, in vitro in the tissues. FASEB J 6: 2639–45

Seeger MA, Haffley L, Kauffman TC (1988) Characterization of *amalgam*: a member of the immunoglobulin superfamily from *Drosophila*. Cell 55: 589–600

Shapiro L, Fannon AM, Kwong PD, Thompson A, Lehmann MS, Grubel G, Legrand JF, Als-Nielson J, Colman DR, Hendrickson WA (1995a) Structural basis of cell-cell adhesion by cadherins. Nature 374: 327–337

Shapiro L, Kwong PD, Fannon AM, Colman DR, Hendrickson WA (1995b) Considerations of the folding topology and evolutionary origin of the cadherin domains. Proc Natl Acad Sci USA 92: 6793–6997

Shaw A, Fortes PAG, Stout CD, Vacquier VD (1995) Crystal structure and subunit dynamics of the abalone sperm lysin dimer: egg envelope dissociates dimers, the monomer is the active species. J Cell Biol 130: 1117–1125

Sim BKL, Chitnis CE, Wasnioska K, Hadley TJ, Miller LH (1994) Receptor and Ligand domains for invasion of erythrocytes by *Plasmodium falciparum*. Science 264: 1941–1944

Simpson P (1995) Developmental genetics. The Notch connection. Nature 375: 736–737

Siu C-H, Kamboj RK (1990) Cell-cell adhesion and morphogenesis in *Dictyostelium discoideum*. Dev Genet 11: 377–387

Siu C-H, Cho A, Choi AH (1987) The contact site A glycoprotein mediates cell-cell adhesion by homophilic binding in *Dictyostelium* discoideum. J Cell Biol 105: 2523–2533

Smith JD, Chitnis CE, Craig AG, Roberts DJ, Hudson-Taylor DE, Peterson DS, Pinches R, Newbold CI, Miller LI (1995) Switches in expression of *plasmodium falciparum var* genes correlate with changes in antigenic and cytoadherent phenotypes of infected erythrocytes. Cell 82: 101–110

Snow PM, Bieber AJ, Goodman CS (1989) Fasciclin II: a novel homophilic adhesion molecule in *Drosophila*. Cell 59: 313–323

Spillmann D, Thomas-Oates JE, van Kuik JA, Vliegenthart JF, Misevic G, Burger MM, Finne J (1995) Characterization of a novel sulfated carbohydrate unit implicated in the carbohydrate-mediated cell aggregation of the marine sponge microciona prolifera. J Biol Chem 270: 5089–5097

Sprague GF (1991) Genetic exchange between kingdoms. Curr Opin Genet Devel 1: 530–533

Springer TA (1990) Adhesion receptors of the immune system. Nature 346: 425–433

Springer WR, Cooper DNW, Barondes SH (1984) *Discoidin I* is implicated in cell-substratum attachment and ordered cell migration of *Dictyostelium discoideum* and resemble fibronectin. Cell 39: 557–564

Steinemann C, Hintermann R, Parish RW (1979) Identification of a developmentally regulated plasma membrane glycoprotein involved in adhesion of *Polysphondylium pallidum* cells. FEBS Lett 108: 379–385

Stern MJ, DeVore DL (1994) Extending and connecting signaling pathways in *C. elegans*. Dev Biol 166: 443–459

Su X-Z, Heatwole VM, Wertheimer SP, Guinet F, Herrfeldt JA, Peterson DS, Ravetch JA, Wellems TE (1995) The large diverse family *var* encodes proteins involved in cytoadherence and antigenic variation of *Plasmodium falciparum*-infected erythrocytes. Cell 82: 89–100

Takeichi M, Atsumi T, Yoshida C, Uno K, Okada TS (1981) Selective adhesion of embryonal carcinoma cells and differentiated cells by Ca2+-dependent sites. Dev Biol 87: 340–350

Teunissen AW, Holub E, van der Hucht J, van den Berg JA, Steensma HY (1993) Sequence of the open reading frame of the *FLO1* gene from *Saccharomyces cerevisiae*. Yeast 9: 423–427

Toda K, Bozzaro S, Lottspeich F, Merkl R, Gerisch G (1984) Monoclonal anti-glycoprotein antibody that blocks cell adhesion in *Polysphondylium pallidum*. Eur J Biochem 140: 73–81

Tomasiewicz H, Ono K, Yee D, Thompson C, Goridis C, Rutishauser U, Magnuson T (1993) Genetic deletion of a neural cell adhesion variant (N-CAM-180) produces distinct defects in the central nervous system. Neuron 11: 1163-1174

Tomson AM, Demets R, Musgrave A, Kooijman R, Stegwee D, van den Ende H (1990) Contact activation in *Chlamydomonas* gametes by increased binding capacity of the sexual agglutinins. J Cell Sci 95: 293–301

Tyler A (1947) An autoantibody concept of cell structure, growth, and differentiation. Growth 10: 7–19

Van den Vaart JM, Caro LH, Chapman JW, Klis FM, Verrips CT (1995) Identification of three mannoproteins in the cell wall of *Saccharomyces cerevisiae*. J Bacteriol 177: 3104 3110

Vardimon L, Fox LL, Degenstein L, Moscona AA (1988) Cell contacts are required for induction by cortisol of glutamine synthetase gene transcription in the retina. Proc Natl Acad Sci USA 85: 5981–5985

Varner JA, Burger MM, Kaufman JF (1988) Two cell surface proteins bind the sponge *Microciona prolifera* aggregation factor. J Biol Chem 263: 8498–8508

Wainwright PO, Hinkle G, Sogin ML, Stickel SK (1993) Monophyletic origins of the metazoa: an evolutionary link with the fungi. Science 260: 340–342

Weiss P (1947) The problem of specificity in growth and development. J Biol Med Yale 19: 235–278

Weissman G, Riesen W, Davidson S, Waite M (1988) Stimulus-response coupling in marine sponge aggregation: lipid metabolism and the function of exogenously added arachidonic and docosahexanoic acids. Biochim Biophys Acta 960: 351–364

Williams AF, Barclay AN (1988) The immunoglobulin superfamily-domains for cell surface recognition. Annu Rev Immunol 6: 381–405

Wilson HV (1907) Coalescence and regeneration in sponges. J Exp Zool 5: 245–253

Wodarz A, Hinz U, Engelbert M, Knust E (1995) Expression of *Crumbs* confers apical character on plasma membrane domains of ectodermal epithelia in *Drosophila*. Cell 82: 67–76

Wojciechowicz D, Lu C-F, Kurjan J, Lipke PN (1993) Cell surface anchorage and ligand-binding domains of the *Saccharomyces cerevisiae* cell adhesion protein α-agglutinin, a member of the immunoglobulin superfamily. Mol Cell Biol 13: 2554–2563

Wyss DF, Choi JS, Li J, Knoppers MH, Willis J, Arulanandam ARN, Smolyar A, Reinherz EL, Wagner G (1995) Conformation and function of the N-linked glycan in the adhesion domain of human CD2. Science 269: 1273–1278

Yoshida M (1987) Identification of carbohydrate moieties involved in EDTA-stable or EDTA-sensitive cell contact of *Dictyostelium* discoideum. J Biochem (Tokyo) 101: 1233–1245

Yoshida M, Matsui T, Fuse G, Ochai S (1993) Carbohydrate structures of the cell adhesion molecule, contact site A, from *Dictystelium* discoideum. FEBS Lett 318: 305–309

Ziegler U, Stidwill RP (1992) The attachment of nematocytes from the primitive invertebrate *Hydra* to fibronectin is specific and RDG-dependent. Exp Cell Res 202: 281–286

Zusman S, Grinblat Y, Yee G, Kafotis FC, Hynes RO (1993) Analysis of PS integrin function during *Drosophila* development. Development 118: 737–750

Animal Lectins as Cell Surface Receptors: Current Status for Invertebrate Species

G.R. Vasta and H. Ahmed[1]

1 Introduction

Although a considerable body of information about the structural, functional and genetic aspects of cell surface lectins from vertebrates, such as asialoglycoprotein liver receptors, mannose-6-phosphate receptors, selectins and membrane mannose binding proteins, is currently available, equivalent research initiatives are yet to be accomplished in comparable depth for invertebrate species. Moreover, despite the fact that the first animal lectin, incidentally from the albumin gland of the snail *Helix pomatia*, was described almost a century ago (Camus 1899), the experimental evidence needed for understanding the biological functions of these sugar-binding proteins, was lacking until the past two decades. Lectins can recognize endogenous ligands (self-recognition) or exogenous ligands (nonself-recognition) and it is now widely accepted that several biological processes based on cell-cell or cell-extracellular matrix interactions, such as tissue organization and development, glycoprotein catabolism, inflammation, fertilization and microbial attachment, are mediated, at least in part, by lectins that are present on the cell surface as integral or adsorbed components for both vertebrate and invertebrate species (Drickamer and Taylor 1993). Modified endogenous structures, such as unusual carbohydrate moieties present on tumors or cells infected by pathogens (virus or bacteria), may also be recognized as nonself by lectins. In addition to acting as recognition molecules, some humoral lectins, such as the mannose-binding receptor described in detail in mammals, function as effector factors by promoting phagocytosis and activating the complement system (Lu et al. 1990).

Because invertebrates lack the characteristic immunoglobulin-mediated immunity of vertebrates, humoral and cell-associated lectins and immunoglobulin superfamily members, such as hemolin, have been proposed to be involved in host defense mechanisms as recognition factors (Vasta 1991). Like their vertebrate equivalents, it has been recently shown that certain lectins from invertebrate species can function not only as recognition but also as

[1]Center of Marine Biotechnology, University of Maryland Biotechnology Institute, 701 East Pratt Street, Baltimore, MD 21202, USA

effector factors exhibiting antibacterial activity (Saito et al. 1995) or triggering melanization (Chen et al. 1995). Putative homologues of the MBPs have been described in invertebrates (Giga et al. 1987; Muramoto et al. 1985; Takahashi et al. 1985; Muramoto and Kamiya 1990; Suzuki et al. 1990; Himeshima et al. 1994; Vasta et al. 1996) and it has been proposed that these, together with the pentraxins, such as C-reactive protein (CRP) and serum amyloid P (SAP), and immunoglobulin superfamily members constitute ancient recognition/effector defense system (Vasta 1991). Therefore, it is becoming increasingly evident that, in addition to the various aforementioned functions, such as cell-cell and cell-extracellular matrix interactions, glycoprotein trafficking and transmembrane signal transduction, both humoral and cell surface lectins constitute a recognition/effector defense system that has been conserved through those lineages that gave origin to the chordates and is still operative in mammals.

In this review we will briefly describe general aspects of the major animal lectin categories, with particular emphasis on integral membrane lectins and their roles as cell surface receptors, to later focus on the most recent advances in the identification and characterization of humoral and membrane lectins from invertebrate species.

2 Animal Lectins as Cell Membrane Receptors

As their plant counterparts, most animal lectins are oligomers (although monomeric lectins have been described) of equal or distinct peptide subunits that can be covalently or noncovalently bound to each other. The molecular organization of the subunits may vary on whether they are soluble or integral membrane proteins. The subunits of integral membrane lectins from vertebrates may vary in size, but their molecular structure is common to most transmembrane proteins: a carboxyl-terminal domain, variable in length and located in the extracellular side of the membrane that carries the carbohydrate recognition domain and putative glycosylation sites, followed by a short transmembrane hydrophobic region and finally an intracytoplasmic amino-terminal segment variable in length and structure that may exhibit putative phosphorylation sites. This is the general pattern of the type II integral membrane lectins. In the less frequent type I the overall orientation of the polypeptide chain is reversed, with the carboxyl-terminal region of the molecule constituting the intracytoplasmic domain. Soluble lectin subunits differ from integral membrane lectins in that hydrophobic residues are distributed throughout the polypeptide sequence without forming a transmembrane domain. The carboxyl-terminal region usually contains the carbohydrate recognition domain, whereas remaining portions of the molecule may exhibit domains with variable structure. These consist of (1) fibrillar collagen-like structures that may activate complement, (2) structures similar to those in the core protein of proteoglycans that interact with glycosaminoglycans, (3) domains similar to epidermal growth factors, or (4) structures similar to those proteins that bind RNA (Drickamer 1988). Among soluble lectins it has been shown that one group, which is ubiquitous in vertebrate tissues and exhibits common properties such as specificity for ß-galactosides and requirement of the presence of free thiol groups, but not divalent cations for binding activity, constitutes a family of proteins which has been substantially conserved in evolution (Barondes 1984). Despite considerable homology, the members of this family may differ in subunit size, architecture, and oligomeric structure (Hirabayashi and Kasai 1993). Based on amino acid sequence similarities, particularly in the carbohydrate binding site, overall domain organization, and properties such as divalent cation dependence and requirement of free thiols, Drickamer (1988) accomplished a major breakthrough in the organization of the field by identifying two major groups of animal lectins: the C- and S- types. In the following years, as more amino acid and nucleotide sequences, and three-dimensional structures become available, animal lectins have been further classified into a number of groups and subgroups that reveal not only the existence of families or at least some structural (and possibly functional) patterns, but also that distinct and probably unrelated, groups of carbohydrate-binding molecules are included under the term "lectin" (Caron et al. 1990; Harrison 1991; Kornfeld 1992; Vasta 1992; Hirabayashi and Kasai 1993; Powell and Varki 1995; Vasta and Ahmed 1995). Some categories, such as the P- and I-type lectins, are composed of only integral membrane proteins (Kornfeld 1992; Powell and Varki 1995), whereas others, such as C-type lectins encompass both soluble and membrane proteins (Drickamer 1988). Among the latter group, those associated with membranes can be integral proteins, such as the asialoglycoprotein receptor or can be adsorbed to the cell surface, in some examples through binding to the glycocalyx.

C-Type Lectins. These proteins comprise a wide variety of soluble and integral membrane proteins that require Ca^{2+} for binding. Carbohydrate binding domains of C-type lectins share a set of invariant and highly conserved amino acids (approximately 15% of the carbohydrate recognition domain), including those cysteine residues involved in disulfide bonds that are required to remain intact for binding activity. This group includes soluble lectins, such as the serum and liver mannose binding proteins, and integral membrane lectins, such as the mammalian asialoglycoprotein receptors, the chicken hepatic lectin, the lymphocyte receptor for the Fc portion of IgE and the mannose receptor from macrophages. C-type lectins may exhibit little or no homology at all in domains other than the carbohydrate recognition site. Those domains confer specific functional properties to each particular lectin, such as hydrophobic domains that anchor the protein to the plasma membrane, complement binding sites, areas for covalent binding of glycosaminoglycans, etc., resulting in structural and functional mosaic or chimeric molecules. C-type lectins have been further classified into four groups, based on their gene organization, e.g., the presence or absence of introns in regions encoding for their carbohydrate recognition domains (CRDs), and primary structure (Bezouska et al. 1991).

Group I includes the cartilage and fibroblast proteoglycan core proteins, with specificities for galactose and fucose. Group II comprises integral membrane proteins such as the hepatic asialoglycoprotein receptors, the macrophage mannose receptor and the IgE-Fc receptor, CD23 (Drickamer and Taylor 1993). These lectins are specific for Gal/Fuc, Gal/GalNAc, or GlcNAc residues. Most of the CRD coding regions characterized for lectins in these first two groups contain introns. Group III includes the so-called "collectins", lectins that are composed of a CRD connected to collagen-like sequences, including the mannose-binding proteins from serum and liver, the pulmonary surfactant protein SP-A, and the bovine conglutinin. Sugar specificities in members of this group include mannose, fucose, GlcNAc, and galactose. Group IV is composed of the membrane cell adhesion molecules known as "selectins", previously known as "LEC-CAMs" (lectin-epidermal growth factor-complement homology-containing cell adhesion molecules), mosaic molecules specific for sialic acids and fucose, among which Mel-14 and LSM-1 (lymphocyte homing receptors or L-selectins), GMP-140 (Platelet granule membrane protein or P-selectin), and ELAM-1(Endothelial cell-leukocyte adhesion molecule or E-selectin) are the best characterized. Genes for lectins included in groups III and IV do not have introns in the CRD coding regions.

Oligomeric membrane receptors consisting of associated polypeptides, each with a single CRD, as in the case of asialoglycoprotein receptors, or those that display multiple CRDs in a single polypeptide, such as in the macrophage mannose receptor, exhibit increased affinity for ligands exhibiting multiple sugar residues (Lee et al. 1984). This "cluster effect" was demonstrated using multivalent synthetic ligands containing terminal galactose or N-acetylgalactosamine and is a feature shared by several C-type lectins such as the chicken hepatic lectin (Lee et al. 1989), asialoglycoprotein receptor (Lee et al. 1984), and the collectins (Drickamer et al. 1986). This phenomenon was confirmed by demonstrating that isolated CRDs obtained by proteolytic cleavage of the receptor exhibit reduced affinity for a multivalent ligand (Loeb and Drickamer 1988). However, both valency and geometry of the receptors and their carbohydrate ligands would be important determinants of binding affinity (Taylor et al. 1992; Taylor and Drickamer 1993). It has been postulated that the different forms of clustering CRDs as seen in nonovalent or multivalent C-type lectins allow the selection of different types of ligand (Drickamer and Taylor 1993). Clustering of CRDs by association of monovalent subunits, such as in asialoglycoprotein receptors, may result in increased affinity and optimal binding to the multivalent carbohydrate ligands exposed on desialylated complex oligosaccharides from the plasma glycoprotiens cleared by these receptors (Drickamer and Taylor 1993). A linear distribution of CRDs in the macrophage mannose receptor would provide the optimal geometry and therefore increased affinity for the diverse linear carbohydrate structures present in microbial surfaces. Moreover, the presence of multiple CRDs in the mannose receptor may provide the flexibility needed for its interaction with a variety of foreign ligands (Drickamer and Taylor 1993).

C-type lectins that are true integral membrane proteins have been proposed to function as membrane receptors for endogenous or exogenous ligands. Endogenous ligands would include those intact or altered carbohydrate moieties present in humoral glycoproteins or surfaces of normal, tumor, and pathogen-infected cells within the organism. Under exogenous ligands we consider those carbohydrate moieties present in foreign substances, including the cell surfaces of putative microbial pathogens. Among the receptors for endogenous ligands a number of type II integral membrane proteins with similar structures have been characterized in hepatocytes, macrophages, and lymphoid cells, and include the mammalian asialoglycoprotein receptor, chicken hepatic lectin, macrophage galactose receptor, and several others (reviewed in Drickamer and Taylor 1993). The hepatocyte asialoglycoprotein receptors are membrane lectins that mediate the clearance of "aged" desialylated plasma glycoproteins through binding to the subterminal galactose or N-acetylgalactosamine, in mammals (reviewed in Ashwell and Harford 1982; Spiess 1990), or GlcNAc in birds (Kawasaki and Ashwell 1977). The endocytosed glycoproteins are catabolized or re-sialylated and returned to circulation. Therefore, the cleavage of terminal sialic acids by a putative plasma neuraminidase, with subsequent exposure of masked carbohydrate moieties, would determine the half-life of plasma glycoproteins in circulation (Drickamer 1991). The mammalian asialoglycoprotein receptor is a hetero-oligomer consisting of two distinct subunits, each capable of binding asialoglycoproteins independently (Rice et al. 1990), but both subunits are required at the cell surface for internalization of the ligand (McPhaul and Berg 1986). This receptor has been purified as a hexamer of subunits that interact through the transmembrane domains, but other stoichiometries may be present in the plasma membrane. In contrast, and although the overall structure of the subunits of the chicken hepatic lectin is very similar to the asialoglycoprotein receptor, the avian counterpart is a homotrimer (Verrey and Drickamer 1993). A homo-ologomeric galactose-specific receptor, structurally similar to the mammalian asialoglycoprotein receptor, has been identified in peritoneal macrophages (Ii et al. 1990) and may participate in tumor surveillance (Sato et al. 1992). Likewise, the Kupffer cell receptor, a fucose- and galactose-binding lectin (Hoyle and Hill 1988) may recognize tumor cells (Vavasseur et al. 1990) through glycolipid moieties (Tiemeyer et al. 1992). Although it is similar to those described above, the extracytoplasmic region is longer and more complex (Beavil et al. 1992).

In addition to macrophages, lymphocytes express various cell surface carbohydrate-binding receptors with C-type CRDs for endogenous ligands. Some receptors belong to the type II transmembrane proteins and are overall similar to the ones described for phagocytic cells, although their function(s) related to internal defense mechanisms as well may be considerably different. CD23 is characterized as a B-cell low-affinity and Ca^{2+}-dependent IgE receptor (Ludin et al. 1987). The ligand on the IgE is likely on the Fc portion but its nature remains unclear (Bettler et al. 1989, 1992; Vercelli et al. 1989; Richards and Katz 1990). A second ligand for CD23 has been identified as CD21 (Aubry et al. 1992; Pochon et al. 1992) and the interaction would be mediated by fucose 1-phosphate residues on the latter and complexed IgE (Gordon 1994).

Although it is currently accepted that recognition mechanisms by natural killer cells for their target cells are different from those that involve rearrangement mechanisms and T-cell receptors (Shinkai et al. 1992), thymocytes and T-cells may express other constitutive or inducible cell surface receptors, such as CD69 (Ziegler et al. 1994), encoded by multigene families originally described on natural killer cells. In natural killer cells, these type II receptors have distal extracytoplasmic C-type, Ca^{2+}-dependent CRDs and include the genetically linked Ly-49 (mouse) (Yokoyama et al. 1989; Wong et al. 1991; Brennan et al. 1994), NKR-P1 (mouse, rat, and human; Yokoyama et al. 1991), and NKG2 (human; Houchins et al. 1991) multigene families. Within each multigene family, members exhibit high similarity throughout the primary structure, but only residues present in the CRD (about 25% identity) are shared between families (Houchins et al. 1991; Wong et al. 1991; Yokoyama et al. 1991). In the mouse, Ly-49 (Brennan et al. 1995) represents a multigene family composed of at least eight distinct genes and members such as Ly-49A and Ly-49C can be expressed by distinct natural killer cell subsets. Target cells are recognized through surface class I major histocompatibility complex antigens (Brennan et al. 1994; Daniels et al. 1994). The inhibitory properties of sulfated glycans for Ly-49C-mediated cell adhesion suggest that recognition of target cells by this natural killer cell receptor is mediated by protein-carbohydrate interactions.

Both lymphocytes and circulating phagocytic cells, such as neutrophils and monocytes, together with platelets and endothelial cells, express another class of cell surface C-type receptors for endogenous ligands, currently known as selectins (Bevilacqua 1993). The cell surface ligands recognized, such as the sialyl Lewis X and related structures, contain either sialic acids or sulfate groups, and fucose (Aruffo et al. 1991; Lasky 1992; Yuen et al. 1992). Unlike the type II membrane receptors described above, selectins are type I transmembrane proteins with C-type, Ca^{2+}-dependent CRDs located on the extracytoplasmic amino-terminus of the polypeptide (Watson et al. 1990; Kansas et al. 1991), which include additional regions such as epidermal growth factor-like and complement-binding domains. These receptors mediate the rather weak initial phase of adhesion between T-cells, neutrophils, monocytes and platelets, and endothelia during inflammatory processes (Lasky 1992). The T-cell homing receptor or L-selectin is involved in the settlement of these cells in peripheral lymph nodes. P-selectins (formerly known as GMP140) are expressed on the surface of platelets and mediate their binding to endothelial cells. Likewise, the endothelial cells express similar receptors, E- and P-selectins, that interact with circulating phagocytic cells in vascular spaces, such as monocytes and neutrophils, facilitating their adhesion and subsequent diapedesis (Lasky 1992; Bevilaqua 1993).

In addition to binding endogenous glycoconjugates such as lysosomal enzymes (Stahl and Schlesinger 1980) and tissue plasminogen activator (Otter et al. 1991) released during the acute phase response, the mannose-binding receptors of macrophages and hepatic endothelial cells can also directly recognize exogenous ligands such as carbohydrate moieties present on microbes. These self/nonself-recognition endocytic receptors mediate the phagocytosis of viral, bacterial, fungal, or protozoan pathogens (Chakraborty and Das 1988; Ezekowitz et al. 1990, 1991) through exposed carbohydrates such as mannose, fucose, and N-acetylglucosamine that are not present in the host normal tissues or soluble glycoconjugates, and therefore, play a role in internal defense (Stahl 1990). Like the selectins, the mannose receptor is a type I transmembrane protein (Taylor et al. 1990), but the presence of multiple CRDs in the polypeptide represent a unique feature within the C-type lectins, since the "cluster effect", i.e., the increased affinity for multivalent ligands, is achieved without the association of multiple subunits. In addition to eight CRDs, the amino-terminal portion of the polypeptide contains a cysteine-rich domain and a fibronectin type II repeat of unknown function, which is not required for internalization of glycans (Taylor et al. 1992; Taylor and Drickamer 1993). Another membrane lectin receptor, possibly with analogous functions in nonself recognition, has been identified in placental tissues (Curtis et al. 1992). However, unlike the macrophage mannose receptor, this lectin is a type II transmembrane protein. This receptor mediates, in vitro, the internalization of HIV through binding to mannose present on the envelope glycoprotein gp120, possibly reflecting its bioloical role (Curtis et al. 1992).

Galectins (S-Type Lectins). Galectins (Barondes et al. 1994), also known as "soluble ß-galactoside-binding lectins" (Barondes 1984), "S-type lectins" (Drickamer 1988), and "galaptins" (Harrison and Chesterton 1980) bind ß-galactosyl residues and require the presence of free thiols but not Ca^{2+} or other divalent cations, for binding activity. Their location is mostly intracellular in the cytoplasmic compartment. Therefore, no detergents are required for their extraction and hence are considered soluble lectins, although they may be present on plasma membrane, in some cases associated through carbohydrate binding sites. These lectins exhibit considerable similarities in the primary structure and, as the C-type lectins, exhibit an invariant residue pattern in the carbohydrate recognition domains (Drickamer 1988). The conserved residues are different from the C-type lectins and do not include cysteines (Liao et al. 1994). Galectin subunits vary in size (Harrison 1991) and domain organization (Hirabayashi and Kasai 1993), and may be associated through noncovalent interactions as homo-dimers (Kasai 1990; Lotan 1992). The galectin from electric eel (electrolectin) is a dimer containing identical 16-kDa subunits (Paroutaud et al. 1987), whereas that from conger eel skin mucus contains two 14-kDa subunits (Muramoto and Kamiya 1992). Based on the primary structure and polypeptide architecture of the subunits, galectins have been now classified into four main groups (Barondes et al. 1994). The "galectin-1" group includes homodimeric species of approximately 14.5 kDa subunit size, such as the human placenta (Hirabayashi and Kasai 1988), human lung (Hirabayashi et al. 1989), rat lung (Clerch et al. 1988), and bovine heart (Abbott et al. 1989) lectins. The term "galectin-2" has been reserved for a distinct type of homodimeric soluble lectin like the human 14 kDa lectin-II (Gitt and Barondes 1991). All monomeric soluble lectins of 26.2–30.3 kDa (e.g., CBP-35, Jia and Wang 1988; Mac-2, Cherayil et al. 1990; IgEBP, Robertson et al. 1990; L-34, Raz et al. 1989) are now included in the "galectin-3" group. Finally, the "galectin-4" class includes soluble lectins that contain two carbohydrate-binding domains in a single polypeptide chain, such as those previously described as L-36 and RIH (Oda et al. 1993). "Galectin-1" and "galectin-2" groups represent the first group of an earlier classification by Harrison (1991) and part of the "proto" type defined by Hirabayashi and Kasai (1993). "Galectin-3" is equivalent to the third group of Harrison's classification and to the "chimera" type (Hirabayashi and Kasai 1993). "Galectin-4" represents the "tandem-repeat" type (Hirabayashi and Kasai 1993) and would correspond to the third group of Harrison's classification. The positions of mammalian soluble lectins of 16–22 kDa (second group of Harrison's classification) and those from lower vertebrates, such as *Xenopus laevis* (Marschal et al. 1992), and invertebrate species, such as the nematode *Caenorhabditis elegans* ("tandem-repeat" type; Hirabayashi et al. 1992) and the sponge *Geodia cydonium* ("proto" type; Pfeifer et al. 1993) remain to be settled in the latest classification (Barondes et al. 1994).

Although galectins have preferential affinities for lactose, N-acetyllactosamine and Galß1, 3GlcNAc over the terminal monosaccharide (D-galactose), from the current knowledge of their fine carbohydrate-binding specificities it is clear that there are two main types. The first type corresponds to those soluble lectins like the human lung 14 kDa (Sparrow et al. 1987), human spleen (Ahmed et al. 1990), rat lung 14 kDa (Leffler and Barondes 1986), and bovine spleen (Ahmed et al. 1994) lectins that have very strict requirements for the orientation or substituents on the 3-OH of Glc/GlcNAc (in Galß1, 4Glc or Galß1, 4GlcNAc) or the 4-OH of GlcNAc (in Galß1, 3GlcNAc). Unlike the first type, the rat intestine domain I, the *Xenopus laevis* (Marschal et al. 1992) and the sponge *Geodia cydonium* SP-I (Hanisch et al. 1984) lectins show decreased binding with N-acetyllactosamine. For the rat intestine domain II (Oda et al. 1993), rat lung 29 kDa (Leffler and Barondes 1986) and human lung 30 kDa (Sparrow et al. 1987) lectins, the binding activities are significantly increased with lactose substituted at both 2'-OH and 3'-OH of the Gal residue (GalNAcα1, 3[Fucα1, 2]Galß1, 4Glc).

The three dimensional structures of members of this lectin family have been resolved recently, providing new information concerning not only the involvement of particular amino acid residues in binding with hydroxyl groups of their carbohydrate ligands, but also on those interactions that determine the architecture of the binding site (Lobsanov et al. 1993; Bourne et al. 1994; Liao et al. 1994; Bianchet et al. 1996). We have cocrystallized the galectin-1 from the bovine spleen with its preferred ligand, N-acetyllactosamine, and determied the 3-D structure and nature of the interactions involved (Liao et al. 1994). Arg^{48}

and His^{44} provide the main interaction with the OH on C4 of galactose residue. The OH on C6 of galactose interacts with Asn^{61}, whereas the OH of C3 of the GlcNAc residue interacts with Arg^{48}, Glu^{71}, and Arg^{73}. The axial orientation of the OH at C4 is important and is restricted by Trp^{68}. The residues Arg^{48}, Asp^{54}, Glu^{71}, and Arg^{73} interact with each other through a network that provides the correct architecture of the binding site. The resolution of the structures of the galectin from ovary of amphibian *Bufo arenarum* complexed with two different sugars, N-acetyllactosamine and thiodigalactoside, provided us further information about the interaction of a carbohydrate-recognition domain with different ligands (Bianchet et al. 1996). The structures of human recombinant galectin-2 and the galectin-1 from bovine heart are also available now (Lobsanov et al. 1993; Bourne et al. 1994). The cross-linking of the bovine heart galectin-1 by a biantennary N-acetyllactosamine-containing oligosaccharide as shown by 3-D structure (Bourne et al. 1994) may reflect interactions that occur when galectins are present on the cell surface. S-type lectins from different species are now crystallized in our laboratory and the resolution of these structures will allow a full understanding of the architecture, function, and evolution of their carbohydrate-binding sites.

Based on this information, we have compared the primary structure of the galectins sequenced so far. All the amino acids (His^{44}, Arg^{48}, His^{52}, Asp^{54}, Asn^{61}, Trp^{68}, Glu^{71}, and Arg^{73}), identified above as participating in the binding, are invariant in the sequences of human placenta, human lung 14-kDa type I, bovine heart, bovine spleen, bovine fibroblast, rat lung 14 kDa, mouse 14 kDa chicken 14 kDa, chicken 16 kDa, and electric eel galectins. The fine carbohydrate-binding specificities of most of them are very similar presumably because of extensive identities in these amino acids. The carbohydrate recognition domain (CRD) of these soluble lectins can be considered invariant and we proposed that this be called a "type I" (or "conserved") CRD (Ahmed and Vasta 1994a). For all other galectins sequenced so far, His^{44}, Arg^{48}, Asn^{61}, Trp^{68}, and Glu^{71} are also invariant, but His^{52}, Asp^{54}, and Arg^{73} can be substituted with a different amino acid, or either Asp^{54} or both His^{52} and Asp^{54} can be deleted. Since unlike the "type I" CRDs described before, the primary structures of these lectins' CRDs are variable, especially between the positions 50–56 and at position 73, and therefore their carbohydrate recognition properties are different, we propose these be called "type II" (or "variable") CRDs. Within this category we include the rat intestine 36-kDa domain I, rat lung 29 kDa, human lung 30 kDa, clawed frog *Xenopus laevis*, sponge *Geodia cydonium*, mouse 35 kDa, and human 14 kDa type II (Ahmed and Vasta 1994a). Some members of the "proto" type lectins (Hirabayashi and Kasai 1993) and all those included in the "chimera" and "tandem-repeat" types, fall within this "type II" CRD category. The primary structures of human lung 22 kDa, rat lung 18 kDa are not known, but their carbohydrate specificities (Leffler and Barondes 1986) would suggest that their CRDs may also be included in the "type II". From this point of view, the members of the "galectin-1" group would exhibit "type I" CRDs, while those from the "galectin-2", "galectin-3", and "galectin-4" groups would display "type II" CRDs.

Although galectins are not transmembrane proteins, some may be present on the cell surface and have been proposed to mediate cell-cell or cell-extracellular matrix interactions (Mecham et al. 1989; Woo et al. 1990; Zhou and Cummings 1990). In developing muscle tissues, a 14.5-kDa soluble lectin is secreted during differentiation and binds to the laminin present in the basement membrane surrounding each myofiber, inducing loss of cell-substratum adhesion (Cooper et al. 1991). The galectin-1 present on the surface of human melanoma cells participates in adhesion to laminin, modulating the invasion and metastasis (Vandenbrule et al. 1995). The Mac-2 antigen, one of the best-characterized S-type lectins and recently included in the galectin-3 group (Barondes et al. 1994), was found on the surface of murine macrophages (Cherayil et al. 1989). A different tumor cell surface galectin-3 has been shown to be involved in homotypic aggregation of tumor cells in the circulation during metastasis through attachment to the complementary serum glycoproteins, which serves as a cross-linking bridge between adjacent cells (Inohara and Raz 1995).

I-Type Lectins. Recent studies on the membrane proteins CD22 and sialoadhesin have resulted in the identification of a distinct category of lectins (reviewed by Powell and Varki 1995). A unique feature of this group resides in their membership to the immunoglobulin superfamily exhibiting a characteristic VI-C2n domain structure (Crocker et al. 1994). All members of this group are transmembrane proteins, some with large cytosolic domains with multiple phosphorylation sites, and several extracellular domains. CD22, a cell surface phosphoglycoprotein present on resting mature B-cells, has seven extracellular domains with the binding region in the first two domains (Engel et al. 1995). The role of sialic acids in the moieties recognized by the receptor has been demonstrated by the use of sialidases, sialyltransferases, and mild periodate oxidation (Powell et al. 1993; Sgroi et al. 1993). This receptor would facilitate antigen-dependent B-cell triggering by association with the B-cell antigen receptor and with cytoplasmic tyrosine kinases (Schulte et al. 1992; Leprince et al. 1993), but it may also recognize ligands on activated lymphocytes, monocytes, and endothelial cells promoting intercellular adhesion (Engel et al. 1993). Another I-type lectin present on selected populations of bone marrow and tissue macrophages, sialoadhesin, exhibits 17 extracellular Ig-like domains (Crocker et al. 1994), the first four NH_2-terminal domains with considerable homology with CD22 (Crocker et al 1994) and proposed to constitute the binding site mediating adhesion of the macrophages to various lymphohematopoietic cells in a sialic acid-dependent manner (Crocker et al. 1994). Additional proteins characterized in other contexts, such as the human myelin-associated glycoprotein, the early myeloid and monocyte marker CD33, and the avian Schwann cell myelin protein, all of unknown function(s), exhibit considerable sequence homology to CD22 and sialoadhesin (Freeman et al. 1995). Based on these structural similarities, further studies aimed at assessing their possible sialic acid-dependent recognition properties, yielded positive results (Freeman et al. 1995). In each case the first two amino-terminal Ig-like domains (V-and C2-type, respectively), which constitute the entire extracelluar portion in CD33, appear to be necessary and sufficient for sialic acid-dependent binding. Regions of highly conserved residues have been identified in members of the I-lectin group, which are not shared with other immunoglobulin superfamily members that map to interstrand regions that form the antigen-binding site in classical immunoglobulins. Other IgSF members such as the cell adhesion molecules NCAM and ICAM-1 may have lectin-like properties that are independent of their homotypic protein-based interactions. The neural cell adhesion molecule (NCAM) can recognize high mannose oligosaccharides on other protein and heparin, whereas ICAM-1 can bind hyaluronan and leukosialin, a sialylated glycoprotein.

Pentraxins. Although the vertebrate pentraxins, such as C-reactive protein, serum amyloid P, and serum amyloid A, constitute a well-defined family of proteins, their lectin-like properties, i.e., binding to carbohydrates and related structures, divalent cation dependence, biological functions and overall molecular structure (Baltz et al. 1982; Uhlenbruck et al. 1982; Hind et al. 1985; Kilpatrik and Volanakis 1985; Tennent and Pepys 1994) have suggested that these molecules may well be incorporated as an additional group of animal lectins (Vasta 1990; Powell and Varki 1995). Like members of the collectin family, some pentraxins behave as acute phase reactants, rapidly increasing their plasma concentration up to 1000-fold or more in response to stress, injury, or infection (Liu et al. 1982). However, in some species, i.e., plaice, pentraxins are constitutive proteins present at high concentrations in the normal plasma (Baltz et al. 1982). The term "pentraxin" derives from their molecular structure: cyclic pentamers of identical subunits with M_r ranging from 20000 to 30000, noncovalently bound in most examples (de Beer et al. 1982). A member of this family with hexameric structure, characterized in the horseshoe crab *Limulus polyphemus*, is an exception (Marchalonis and Edelman 1968). Pentraxins constitute a highly conserved family and distant species exhibit a considerable level of structural similarity. For example, CRPs that have been isolated from plaice and human exhibit 40% homology in their amino acid sequence (Baltz et al. 1982). CRPs share common features independently of the source species, such as the size of the subunits (20 000 – 30 000) and the presence of Ca^{2+} -dependent phosphocholine-binding sites. Other properties such as glycosylation and levels in normal plasma are variable. Pentraxins share some properties with lectins. In addition to the PC-binding properties, CRP binds and precipitates galactins (Baltz et al. 1982; Uhlenbruck et al. 1982), fungal extracts (Baldo et al. 1977), and carageean gums (Liu et al. 1982) in a Ca^{2+}-dependent manner. SAP, which binds the pyruvate acetal of galactose, has been considered a vertebrate serum lectin that modulates immune responses (Linn et al. 1984).

The biological roles of most pentraxins are not yet fully understood. The fact that their ligands phosphocholine and galactose derivatives are widely distributed in microorganisms, fungi, and parasites suggests that they may be related to self/nonself discrimination mechanisms. Most pentraxins characterized so far are synthesized in the liver under regulation by cytokines such as IL-6 and TNF, and secreted to plasma. However, some members of this group have been identified in association with cell surfaces of lymphoid cells and macrophages. C-reactive protein has been identified as a galactose-specific particle receptor on liver macrophages and its properties are certainly analogous to those macrophage lectin receptors described above. It has been shown that CRP acts as an antimicrobial opsonin by activating complement and promoting phagocytosis (Hokoma et al. 1962). CRP-ligand complexes activate complement via a classical pathway and interact with neutrophils to generate the phagocytic signal (Kilpatrick and Volanakis 1985). Both mammalian and *Limulus polyphemus* CRP can induce platelet-mediated cytotoxic activity against parasites such as *Schistosoma* sp. (Bout et al. 1986). In addition, CRP also interacts strongly with chromatin and it has been hypothesized that another function of CRP is to protect against chromatin release to the circulation by cell damage (Robey et al. 1984).

P-Type Lectins (Mannose 6-Phosphate Receptors). Mannose-6-phosphate receptors (P-lectins or MPRs) are type I trasmembrane proteins present in the Golgi and plasma membrane with roles in intracellular targeting of lysosomal enzymes and transmembrane signal transduction (reviewed by Kornfeld 1992). Both lysosomal enzymes (acid hydrolases) and secretory proteins are synthesized in the rough endoplasmic reticulum and posttranslationally glycosylated. However, the lysosomal hydrolases are phosphomannosylated and therefore, recognized by the MPRs in the Golgi. The complexes are transported to the acidic endosomes, dissociated and the receptor recycled to the Golgi or to the plasma membrane (Kornfeld 1992). Two different but structurally related mannose-6-phosphate receptors (MRPs) have been described: both mediate the trafficking of acid hydrolases to lysosomes but only the cation-independent mannose-6-phosphate/insulin-like growth factor II receptor [(CI)MPR] binds IGF-II, whereas the second, a cation-dependent receptor [(CD)MPR] in bovine and rat, does not (Kornfeld 1992). In some cell types, the mannose-6-phosphate/IGF-II receptor may also be involved in transmembrane signal transduction and it has been postulated that it may mediate hormone clearance and activation, and turnover of extracellular matrix proteoglycans (Kornfeld 1992). The oligomeric structure of the MPRs is unclear. MPRs have been isolated as monomers, dimers, and tetramers that may reflect subunit associations that vary with the cellular compartment. MPRs exhibit a signal sequence and extracytoplasmic, transmembrane, and intracytoplasmic domains. The CRD consists of a similar sequence motif of about 147 amino acid residues, present in the extracytoplasmic domains of both (CI)MPR and (CD)MPR, although 15 repeats have been found in the

(CI)MPR (Kornfeld 1992). This P-type CRD motif is unique among animal lectins and, when interacting with multivalent ligands, may exhibit a "cluster effect" similar to that described above for the C-type lectins.

Heparin-Binding Lectins. A number of reports have described proteins, such as neural cell adhesion molecule, fibronectin endothelial cell growth factors (reviewed in Kjellen and Lindahl 1991) that bind to heparin and other glycosaminoglycans. Although some of these have been identified as members of the well-established families of proteins, other have been grouped as a separate category of lectins (Roberson et al. 1981; Eloumami et al. 1990) that, together with a number of recently described extracellular matrix-binding proteins such as the hyaluronan-binding proteins, calnexin and calcireticulin, ganglioside-binding proteins and sulfoglucuronosyl lipid-binding protein, would mediate interaction between cells and their environement. Tissues from mammalian nervous systems have been studied in detail and appear to be particularly rich in heparin-binding lectins (Eloumami et al. 1990). In addition, cytokines such as IL-Ia, IL-2, and TNF exhibit lectin properties (Muchmore and Decker 1987; Sherblom et al. 1988, 1989) but have not been classified within the C-, S-, P-, or I-type lectin categories indicated above.

3 Lectin Families in Invertebrate and Protochordate Species. Their Association with the Hemocyte Plasma Membrane

Within the invertebrate and protochordate taxa, lectins related to the major groups of vertebrate lectins described above have been identified. A number of invertebrate lectins that show significant homology to membrane or soluble C-type lectins from vertebrates have been isolated and characterized. This group includes lectins from the flesh fly *Sarcophaga peregrina* larva (Takahashi et al. 1985), the acorn barnacle *Megabalanus rosa* (Muramoto et al. 1985; Muramoto and Kamiya 1990), the sea urchin *Anthocidaris crassispina* (Giga et al. 1987), the sea cucumber *Stichopus japonicus* (Himeshima et al. 1994), and the tunicates *Polyandrocarpa misakiensis* (Suzuki et al. 1990) and *Clavelina picta* (Vasta et al. 1996). Like most C-type lectins, these proteins share a carboxyl-terminal domain that contains the CRD with a highly conserved set of residues that include cystein and tryptophan. In echinoidin, a lectin from coelomic fluid of the sea urchin *Anthocidaris crassispina*, the carboxyl-terminal half is similar to the mannose-binding lectins A and C from rat liver (35 and 32% identity, respectively), the chicken hepatic lectin, and the rat asialoglycoprotein receptor (Giga et al. 1987). This would suggest that the carboxyl-terminal region of the molecule containing the carbohydrate-binding site is an early development in the evolution of animal lectins (Giga et al. 1987). It also shows homology to the central portion of the lectin from the fly *Sarcophaga peregrina* (28% identity; Takahashi et al. 1985) which in turn shows 18% identity with the rat mannose-binding proteins (Drickamer et al. 1986). The galactose-binding lectin from another echinoderm, the sea cucumber *Stichopus japonicus*, is 28.6% identical to echinonectin and exhibits a CRD as well as the Ca^{2+} binding domain on the carboxyl-terminal region of the polypeptide subunit (Himeshima et al. 1994).

The S-type lectins or galectins are also represented within the invertebrate and protochordate taxa. The galectins from the nematode *Caenorhabditis elegans* (Hirabayashi et al. 1992), the sponge *Geodia cydonium* (Pfeifer et al. 1993), and the tunicate *Clavelina picta* (Ahmed and Vasta 1994b) are included in this group (Hirabayashi and Kasai 1993). Three different subunit sizes (13, 15, and 16 kDa) were found in the galectins of sponge *G. cydonium* (Pfeifer et al. 1993) but only a 32 kDa was present in the nematode *C. elegans* (Hirabayashi et al. 1992). The galectins isolated from the tunicate *C. picta* have subunits of sizes 14.8, 15.8, 33.5, and 37.5 kDa (Ahmed and Vasta 1994b). Within the classification of Hirabayashi and Kasai (1993), the *C. elegans* 32-kDa lectin was included in the "tandem-repeat" type that has only one other member, the 36-kDa lectin from rat intestine (Oda et al. 1993). These lectins have two repeated domains, most likely products of gene duplication, that carry identical or slightly different carbohydrate-binding sites, separated by a region susceptible to proteolytic activity (Leffler et al. 1989). The S-lectin SP-I from the sponge *G. cydonium* was included in the "proto" type (Pfeifer et al. 1993). In the latest classification (Barondes et al. 1994), however, the position of S-lectins from both invertebrate species remains to be settled. With regard to the conservation of amino acid residues in the CRDs, the S-lectins from both *C. elegans* and *G. cydonium* exhibit "type II" (or "variable") CRDs (Ahmed and Vasta 1994a).

Lectins from the horseshoe crab *Limulus polyphemus* (Nguyen et al. 1986a) and the tunicate *Didemnum candidum* (Vasta et al. 1986a) are homologues of vertebrate pentraxins (Vasta 1990). The three subunits of *L. polyphemus* lectin show homology (approximately 25%) to human and rabbit CRP, SAP, and hamster female protein, all mammalian pentraxins that show extensive homology to each other. In addition, the three *L. polyphemus* subunits share an amino-terminal sequence of 44 residues and a carboxyl-terminal sequence from residues 206 to 218 with constant positions for the glycosylation site and six half-cystines that form three disulfide bonds (Nguyen et al. 1986a). *L. polyphemus* lectin (CRP) subunits are encoded by multiple, closely related genes that use the same promoter and start and stop signals, and exhibit extensive homology with each other throughout their nucleotide sequence (Nguyen et al. 1986b). There is evidence that additional homologous genes would code for a series of related proteins, the whole constituting a family of lectins. Human CRP, which is an acute phase protein, exhibits an intron between the signal peptide and the coding sequence containing DNA segments that could potentially adopt the z form. In addition, between the TATA box and the coding region, it contains a *Drosophila* heat shock consensus sequence. Both regions, which have been thought to regulate systhesis of CRP in the mammalian acute phase response, are missing from *L. polyphemus* genes. *L. polyphemus* lectin and human CRP show approximately 25% overall identity in their nucleotide sequences, with two regions of conserved residues. One region (residues 52 to

67) would constitute the binding site for phosphorylcholine and the other (residues 139-153) would bind Ca^{2+} (Nguyen et al. 1986b). Comparisons of amino acid compositions of lectins isolated from invertebrate, protochordate and cyclostome species, with a data base composed of amino acid compositions of 252 proteins (Vasta and Marchalonis 1987), have shown that DCL-I, a lectin isolated from the tunicate *D. candidum* (Vasta and Marchalonis 1986; Vasta et al. 1986a,b), may be structurally related to a mammalian CRP. The comparison of 21 residues of DCL-I N-terminal amino acid sequence with mammalian acute phase proteins is interesting in that several identities are found between DCL-I and CRP and SAP in a region that overlaps with the sequence stretches that exhibit several identities with *L. polyphemus* lectin (Vasta et al. 1986a). Although anti-idiotypic monoclonal antibodies made against the myeloma protein TEPC 15 cross-reacted with *L. polyphemus* lectin and CRP, they did not cross-react with DCL-I (Vasta et al. 1984). This was not surprising since DCL-I binds galactosyl residues (Vasta and Marchalonis 1986) while the other three molecules, TEPC 15, CRP and *L. polyphemus* lectin, bind PC. However, polyclonal antibodies made against CRP cross-react with TEPC-15 and *L. polyphemus* lectin but also with DCL-I (Vasta et al. 1984). This suggests although TEPC-15, CRP, and *L. polyphemus* lectin might share common determinants related to their binding sites for PC, determinants shared by DCL-I and CRP are probably located in other regions of the molecule. Recently, we have shown that polyclonal antibodies against human CRP cross-react with a lectin purified from the tunicate *C. picta*. Most important of all, monoclonal antibodies developed against the purified *C. picta* lectin cross-react with human CRP (Elola and Vasta, in prep.).

Because of the lack of information regarding their primary structures, the vast majority of invertebrate lectins described so far cannot yet be placed in any of the aforementioned groups. Some invertebrate lectins, however, exhibit some features, such as cross-reactivity with certain proteins from vertebrates or specificity for particular ligands, that suggest that they may be related to some of the lectin groups defined for vertebrates species. For example, some lectins from protochordates such as those isolated from the tunicate *Halocynthia pyriformis* show substantial cross-reactivity with antibodies to immunoglobulin polypeptide chains from elasmobranchs (Vasta et al., unpubl). The involvement of anticarbohydrate activity can be ruled out because the tunicate lectins in question are not glycosylated. Therefore, it is possible that these may represent homologues of the I-type lectins or other immunoglobulin superfamily members. A heparin-binding lectin, "anadarin", from the bivalve *Anadara granosa* has been recently isolated and partially characterized (Dam et al. 1994). Similar binding properties have been detected in lectins from the tunicate *Clavelina picta* (Vasta et al., unpubl.), raising the possibility that invertebrates may exhibit lectins related to those heparin/glycosaminoglycan-binding proteins described in vertebrates (Kjellen and Lindahl 1991). The elucidation of the amino acid or nucleotide sequences for those invertebrate lectins will enable us to explain those observations.

Finally, some lectins, such as the L-10 from *Tachypleus tridentatus* (Okino et al. 1995) from which the primary structure is available, show no resemblance to any lectin described so far. As more information on primary structures becomes available in the future, putative relationships with carbohydrate-binding proteins from invertebrates or chordates may emerge. Other carbohydrate-binding proteins such as echinonectin, a galactose-specific lectin from sea urchin embryos that may be analogous to fibronectin (Alliegro et al. 1990; Veno et al. 1990), have not yet been classified within any of the categories indicated above.

Although lectins have been found to be ubiquitous among invertebrate species, only a very small number has been demonstrated as true cell-surface lectins. Different populations of blood cells or hemocytes participate in a series of defense-related reactions including phagocytosis, encapsulation, hemolymph clotting and release of phenoloxidase and melanization in response to challenge with "nonself" components. The participation of lectins as mediators of those immune mechanisms in invertebrates and protochordates is supported by extensive experimental evidence for their roles as agglutinins, opsonins and hemocyte-associated nonself recognition and effector molecules.

True membrane lectins associated with these latter functions can be illustrated by the recently characterized lipopolysaccharide-binding lectin (L6) that has been isolated from detergent extracts of hemocytes from the Japanese horseshoe crab *T. tridentatus* (Saito et al. 1995). This membrane lectin has antibacterial activity, and therefore functions not only as a rec-ognition molecule, but as an effector factor as well. Upon binding of LPS of Gram-negative bacteria, hemocytes begin to degranulate, and the released granular components would initiate hemolymph coagulation, followed by phagocytosis of the invading microbes (Saito et al. 1995).

Many other lectins have been identified, isolated, and cloned from hemocytes in numerous invertebrate species indicating that these proteins are synthesized in those cells. However, the association of the lectin to the plasma membrane has only been demonstrated by indirect methods, such as recognition/endocytosis of complex carbohydrates or cells by hemocytes (Scott 1971; Tyson and Jenkin 1974; Vasta et al. 1982); immunofluorescent staining of intact hemocytes (Amirante and Mazzalai 1978; Vasta et al. 1984a; Lackie and Vasta 1988; Elola and Vasta, unpubl. observ.) or by assessing carbohydrate binding or cell agglutination by microsomal fractions obtained by differential centrifugation of hemocyte extracts (Vasta et al. 1982; Cassels et al. 1986; Elola and Vasta, unpubl.). For example, microsomal fractions of oyster (*Crassostrea virginica*) hemocytes yielded lectin three to four fold higher than the whole hemocyte extracts (Vasta et al. 1982). In the blue crab (*Callinectes sapidus*), lectin-specific activities of the microsomal fractions were 30- to 40-fold higher than the whole hemocyte extracts (Cassels et al. 1986). In many cases, the hemocyte lectin is similar or identical to that present in plasma, suggesting that hemocytes are the source of the humoral lectin (Cassels et al. 1986). Our research on the oyster *Crassostrea virginica* hemocyte and serum lectins showed that the hemocyte membrane lectin shares the carbohydrate specificity of a fraction of the oyster plasma lectins (Vasta et al. 1982). In vivo evidence for the participation of serum and hemocyte-associated lectins in cellular responses to nonself was obtained in the cockroach *Periplaneta americana*: glycoproteins that were the most effective inhibitors for the purified plasma lectins, when injected in solution or bound to Sepharose beads in the hemocoele of the insect, produced a cellular response, 'nodulation, and encapsulation significantly higher than for glycoproteins that behaved as poor inhibitors for the purified lectins (Lackie and Vasta 1988). Members of a lectin family characterized in the same species, however, have been proposed to participate in tissue organization during limb regeneration (Kubo et al. 1993). Analysis of dot and Northern blot hybridization reveals that the gene encoding for hemocytin, a humoral C-type lectin from silkworm (*Bombyx mori*), is transcribed in hemocytes at larval-pupal metamorphosis (Kotani et al. 1995). The plasma lectin CPL-III from the tunicate *Clavelina picta* has been cloned from hemocyte RNA. Monoclonal antibodies specific for this lectin strongly stain the plasma membrane of intact hemocytes and large intracellular granules in permeabilized cells. This lectin recognizes and agglutinates a large variety of environmental bacterial isolates, and binds the structural sulfated glycan of the tunic. Therefore we have proposed that this lectin is involved not only in the recognition, immobilization, and endocytosis of putative environmental microbial pathogens but also in wound repair (Vasta et al., 1996).

4 Summary and Prospects

From the limited information about the primary structure available at the present time, three major groups of humoral and membrane invertebrate lectins can be discriminated: one group would include lectins that show significant homology to membrane-integrated or soluble C-type vertebrate lectins. A second group is homologous to the galectin or S-type lectin family. A third group would be composed of lectins that show homology to vertebrate pentraxins such as C-reactive protein. Additional information from serological cross-reactivities and carbohydrate-binding properties suggest that other lectin groups may be represented in the invertebrate and protochordate taxa. Finally, the primary structure of some invertebrate lectins reveals the lack of similarity to any lectins from vertebrates described so far. If these proteins have no equivalents in extant vertebrate species, they may reflect evolutionary processes divergent or parallel to those that occurred in the chordate lineages. Therefore, it is very likely that careful analysis of the structure/function of humoral and membrane lectins in invertebrates and protochordates will become available in the future, will yield new families of lectins, and will provide information concerning their evolution in invertebrate and protochordate stocks. Clearly, more research is needed for establishing the presence and the nature of the association of lectins in plasma membranes, whether they are integral transmembrane proteins or adsorbed or bound through their ligands to the glycocalyx. Because of their participation in internal defense mechanisms, hemocyte membrane lectins are of particular interest in this regard.

Acknowledgments. Part of the studies described here was supported by grant MCB-91-05875 from the National Science Foundation, grant F31 GM14903 from the National Institutes of Health, DOC Cooperative Agreements #NA47FL-0163 and #NA57FL0039 awarded by NOAA/NMFS, Oyster Disease Research Program and grant #NA90AA-D-SG063 awarded by NOAA through the University of Maryland Sea Grant College to G.R.V.

References

Abbott WM, Mellor A, Edwards Y, Feizi T (1989) Soluble bovine galactoside-binding lectin cDNA reveals the complete amino acid sequence and an antigenic relationship with the major encephalitogenic domain of myelin basic protein. Biochem J 259: 283–290

Ahmed H, Vasta GR (1994a) Galectins: conservation of functionally and structurally relevant amino acid residues defines two types of carbohydrate recognition domains. Glycobiology 4: 545–549

Ahmed H, Vasta GR (1994b) Galectins from vertebrates and invertebrates: comparative studies on physicochemical and carbohydrate-binding properties. Glycobiology 4: 724

Ahmed H, Allen HJ, Sharma A, Matta KL (1990) Human splenic galaptin: carbohydrate-binding specificity and characterization of the combining site. Biochemistry 29: 5315–5319

Ahmed H, Fink NE, Vasta GR (1994) A novel solid-phase assay for lectin binding: comparative studies on ß-galactoside-binding S-type lectins from fish, amphibian, and mammalian tissues. Ann NY Acad Sci 712: 315–317

Alliegro MC, Burdsal CA, McClay DR (1990) In vitro biological activities of echinonectin. Biochemistry 29: 2135–2141

Amirante GA, Mazzalai FG (1978) Synthesis and localisation of hemagglutinins in hemocytes of the cockroach *Leucophaea maderae*. L. Dev Comp Immunol 2: 735–740

Aruffo A, Kolanus W, Walz G, Fredman P, Seed B (1991) CD62/P-selectin recognition of myeloid and tumor cell sulfatides. Cell 67: 35–44

Ashwell G, Harford J (1982) Carbohydrate-specific receptors of the liver. Annu Rev Biochem 51: 531–554

Aubry J-P, Pochon S, Graber P, Jansen KU, Bonnefoy J-Y (1992) CD21 is a ligand for CD23 and regulates IgE production. Nature 358: 505–507

Baldo BA, Fletcher TC, Pepys J (1977) Isolation of a peptido-polysaccharide from the dermatophyte *Epidermophyton floccosum* and a study of its reaction with human C-reactive protein and a mouse anti-phosphorylcholine myeloma serum. Immunology 32: 831–842

Baltz ML, de Beer FC, Feinstein A, Munn EA, Milstein CP, Fletcher TC, March JF, Taylor J, Bruton C, Clarfnp JR, Davies AJS, Pepys MB (1982) Phylogenetic aspects of C-reactive protein and related proteins. Ann NY Acad Sci 389: 49–75

Barondes SH (1984) Soluble lectins: a new class of extracellular proteins. Science 223: 1259–1264

Barondes SH, Castronovo V, Cooper DNW, Cummings RD, Drickamer K, Feizi T, Gitt MA, Hirabayashi J, Hughes C, Kasai K, Leffler H, Liu F, Lotan R, Mercurio AM, Monsigny M, Pillai S, Poirer F, Raz A, Rigby PWJ, Rini JW, Wang JL (1994) Galectins: a family of animal ß-galactoside-binding lectins. Cell 76: 597–598

Beavil AJ, Edmeades RL, Gould HJ, Sutton BJ (1992) α-Helical coiled-coil stalks in the low-affinity receptor for IgE (FceRII/CD23) and related C-type lectins. Proc Natl Acad Sci USA 89: 753–757

Bettler B, Maier R, Ruegg D, Hoffstetter H (1989) Binding site for IgE of the human lymphocyte low-affinity Fce receptor (Fce/CD23) is confined to the domain homologous with animal lectins. Proc Natl Acad Sci USA 86: 7118–7122

Bettler B, Texido G, Raggini S, Ruegg D, Hofstetter H (1992) Immunoglobulin E-binding site in Fce receptor (FceRII/CD23) identified by homolog-scanning mutagenesis. J Biol Chem 267: 185–191

Bevilacqua MP (1993) Endothelial-leukocyte adhesion molecules. Annu Rev Immunol 11: 767–804

Bezouska K, Crichlow GV, Rose JM, Taylor ME, Drickamer K (1991) Evolutionary conservation of intron position in a subfamily of genes encoding carbohydrate-recognition domains. J Biol Chem 266: 11604–11609

Bianchet M, Ahmed H, Vasta GR, Amzel LM (1996) 3-D structure of a soluble ß-galactosyl-binding lectin (galectin) from toad (*Bufo arenarum*) ovary. Nature Struct Biol (Communicated)

Bourne Y, Bolgiano B, Liao D, Strecker G, Cantau P, Herzberg O, Feizi T, Cambillau C (1994) Crosslinking of mammalian lectin (galectin-1) by complex biantennary saccharides. Nature Struct Biol 1: 863–870

Bout D, Joseph M, Pontent M, Vorng HD, Capron A (1986) Rat resistance to schistosomiasis: platelet-mediated cytoxicity induced by C-reactive protein. Science 231: 153–156

Brennan J, Mager D, Jefferies W, Takei F (1994) Expression of different members of the Ly-49 gene family defines natural killer cell subsets and cell adhesion properties. J Exp Med 180: 2287–2295

Brennan J, Takei F, Wong S, Mager DL (1995) Carbohydrate recognition by a natural killer cell receptor, Ly-49C. J Biol Chem 270: 9691–9694

Camus ML (1899) Recherches experimentales sur une agglutinine produite par la glande de l'albumen chez l'*Helix pomatia*. CR Hebd Seances Acad Sci 129: 233

Caron M, Bladier D, Joubert R (1990) Soluble galactose-binding lectins: a protein family with common properties. Int J Biochem 22: 1379–1385

Cassels FJ, Marchalonis JJ, Vasta GR (1986) Heterogenous humoral and hemocyte-associated lectins with N-acylaminosugar specificities from the blue crab, *Callinectes sapidus* Rathbun. Comp Biochem Physiol 85B: 23–30

Chakraborty P, Das PK (1988) Role of mannose/N-acetyglucosamine receptors in blood clearance and cellular attachment of *Leishmania donovani*. Mol Biochem Parasitol 28: 55–62

Chem C, Durrant J, Newton RP, Ratcliffe NA (1995) A study of novel lectins and their involvement in the activation of the prophenoloxidase system in *Blaberus discoidalis*. Biochem J 310: 23–31

Cherayil BJ, Weiner SJ, Pillai S (1989) The Mac-2 antigen is a galactose-specific lectin that binds IgE. J Exp Med 170: 1959–1972

Cherayil BJ, Chaitovitz S, Wong C, Pillai S (1990) Molecular cloning of a human macrophage lectin specific for galactose. Proc Natl Acad Sci USA 87: 7324–7328

Clerch LB, Whitney P, Hass M, Brew K, Miller T, Werner R, Massaro D (1988) Sequence of a full-length cDNA for rat lung ß-galactoside-binding protein: primary and secondary structure of the lectin. Biochemistry 27: 692–699

Cooper DNW, Massa SM, Barondes SH (1991) Endogenous muscle lectin inhibits myoblast adhesion to laminin. J Cell Biol 115: 1437–1448

Crocker PR, Mucklow S, Bouckson V, McWilliam A, Willis AC, Gordon S, Milon G, Kelm S, Bradfield P (1994) Sialoadhesin, a macrophage sialic acid binding receptor for haemopoietic cells with 17 immunoglobulin-like domains. EMBO J 13: 4490–4503

Curtis BM, Scharnowske S, Watson AJ (1992) Sequence and expression of a membrane-associated C-type lectin that exhibits CD4-independent binding of human immunodeficiency virus envelope glycoprotein gp 120. Proc Natl Acad Sci USA 89: 8356–8360

Dam TK, Bandyopadhyay P, Sarkar M, Ghoshal J, Bhattacharya A, Choudhury A (1994) Purification and partial characterization of a heparin-binding lectin from the marine clam *Andara granosa*. Biochem Biophys Res Commun 203: 36–45

Daniels BF, Karlhofer FM, Seaman WE, Yokoyama WM (1994) A natural killer cell receptor specific for a major histocompatibility complex class I molecule. J Exp Med 180: 687–692

de Beer FC, Baltz ML, Munn EA, Feinstein A, Taylor J, Bruton C, Clamp JR, Pepys MB (1982) Isolation and characterization of C-reactive protein and serum amyloid P component in the rat. Immunology 45: 55–70

Drickamer K (1988) Two distinct classes of carbohydrate-recognition domain in animal lectins. J Biol Chem 263: 9557–9560

Drickamer K (1991) Clearing up glycoprotein hormones. Cell 67: 1029–1032

Drickamer K, Taylor ME (1993) Biology of animal lectins. Annu Rev Cell Biol 9: 236–264

Drickamer K, Dordal MS, Reynolds L (1986) Mannose-binding proteins isolated from rat liver contain carbohydrates-recognition domains linked to collagenous tails. Complete primary structures and homology with pulmonary surfactant apoproteins. J Biol Chem 261: 6878–6887

Eloumami H, Bladier D, Caruelle D, Courty J, Jouberet R, Caron M (1990) Soluble heparin-binding lectins form human brain: purification, specificity and relationship to a heparin-binding growth factor. Int J Biochem 22: 539–544

Engel P, Nojima Y, Rothstein D, Zhou LJ, Wilson GL, Kehrl JH, Tedder TF (1993) The same epitope on CD22 of lymphocytes mediates the adhesion of erythocytes, T and B lymphocytes, neutrophils and monocytes. J Immunol 150: 4719–4732

Engel P, Wagner N, Miller AS, Tedder TF (1995) Identification of the ligand-binding domains of CD22, a member of the immunoglobulin superfamily that uniquely binds a sialic acid-dependent ligand. J Exp Med 181: 1581–1586

Ezekowitz RAB, Sastry K, Bailly P, Warner A (1990) Molecular characterization of the human macrophage mannose receptor: demonstration of multiple carbohydrate recognition-like domains and phagocytosis of yeasts in Cos-1 cells. J Exp Med 172: 1758–1794

Ezekowitz RAB, Williams DJ, Koziel H, Armstrong MYK, Warner A, Richards FF, Rose RM (1991) Uptake of *Pneumocystis carinii* mediated by the macrophage mannose receptor. Nature 351: 155–158

Freeman SD, Kelm S, Barber EK, Crocker PR (1995) Characterization of CD33 as a new member of the sialoadhesin family of cellular interaction molecules. Blood 85: 2005–2012

Giga Y, Ikai A, Takahashi K (1987) The complete amino acid sequence of echinoidin, a lectin from the coelomic fluid of the sea urchin *Anthocidaris crassispina*. J Biol Chem 262: 6197–6203

Gitt MA, Barondes SH (1991) Genomic sequence and organization of two members of a human lectin gene family. Biochemistry 30: 82–89

Gordon J (1994) B-cell signalling via the C-type lectins CD23 and CD72. Immunol Today 15: 411–417

Hanisch H-G, Saur A, Müller WEG, Conrad J, Uhlenbruck G (1984) Further characterization of a lectin and its in vivo receptor from *Geodia cydonium*. Biochim Biophys Acta 801: 388–395

Harrison FL (1991) Soluble vertebrate lectins: ubiquitous but inscrutable proteins. J Cell Sci 100: 9–14

Harrison FL, Chesterton CJ (1980) Factors mediating cell-cell recognition and adhesion. Galaptins, a recently discovered class of bridging molecules. FEBS Lett 122: 157–165

Himeshima T, Hatakeyama T, Yamasaki N (1994) Amino acid sequence of a lectin from the sea cucumber, *Stichopus japonicus*, and its structural relationship to the C-type animal lectin family. J Biochem 115: 689–692

Hind CRK, Collins PM, Baltz ML, Pepys MB (1985) Human serum amyloid P component, a circulating lectin with specificity for the cyclic 4,6-pyruvate acetal of galactose. Biochem J 225: 107–111

Hirabayashi J, Kasai K (1988) Complete amino acid sequence of a ß-galactoside-binding lectin from human placenta. J Biochem (Tokyo) 104: 1–4

Hirabayashi J, Kasai K (1993) The faimly of metazoan metal-independent ß-galactoside-binding lectins: structure, function and molecular evolution. Glycobiology 3: 297–304

Hirabayashi J, Ayaki H, Soma G, Kasai K (1989) Cloning and nucleotide sequence of a full length cDNA for human 14kDa ß-galactoside-binding lectin. Biochim Biophys Acta 1008: 85–91

Hirabayashi J, Satoh M, Kasai K (1992) Evidence that *Caenorhabditis elegans* 32 kDa ß-galactoside-binding protein is homologous to vertebrate ß-galactoside-binding lectins. J Biol Chem 267: 15485–15490

Hokoma Y, Coleman MK, Riley RF (1962) In vitro effects of C-reactive proteins on phagocytosis. J Bacteriol 83: 1017–1024

Houchins JP, Yabe T, McSherry C, Bach FH (1991) DNA sequence analysis of NKG2, a family of related cDNA clones encoding type II integral membrane proteins on human natural killer cells. J Exp Med 173: 1017–1020

Hoyle GW, Hill RL (1988) Molecular cloning and sequencing of a cDNA for a carbohydrate binding receptor unique to rat Kupffer cells. J Biol Chem 263: 7487–7492

Ii M, Kurata H, Itoh N, Yamashina I, Kawasaki T (1990) Molecular cloning and sequence analysis of cDNA encoding the macrophage lectin specific for galactose and N-acetylgalactosamine. J Biol Chem 265: 11295–11298

Inohara H, Raz A (1995) Functional evidence that cell surface galectin-3 mediates homotypic cell adhesion. Cancer Res 55: 3267–3277

Jia S, Wang JL (1988) Carbohydrate binding protein 35: complementary DNA sequence reveals homology with proteins of the heterogeneous nuclear RNP. J Biol Chem 263: 6009–6011

Kansas GS, Spertini O, Stoolman LM, Tedder TF (1991) Molecular mapping of functional domains of the leukocyte receptor for endothelium, LAM-1. J Cell Biol 114: 351–358

Kasai K (1990) Biochemical properties of vertebrate 14k ß-galactoside-binding lectins. In: Franz H (ed) Advances in lectin research, vol 3. VEB Verlag Volk und Gesundheit, Berlin, pp 10–35

Kawasaki T, Ashwell G (1977) Isolation and characterization of an avian hepatic binding protein specific for N-acetylglucosamine-terminated glycoproteins. J Biol Chem 252: 6536–6543

Kilpatrik JM, Volanakis JE (1985) Opsonic properties of C-reactive protein. Stimulation by phorbol myristate acetate enables human neutrophils to phagocytize C-reactive protein coated cells. J Immunol 134: 3364–3370

Kjellen L, Lindahl U (1991) Proteoglycans: structures and interactions. Annu Rev Biochem 60: 443–475

Kornfeld S (1992) Structure and function of the mannose 6-phosphate/insulin-like growth factor II receptors. Annu Rev Biochem 61: 307–330

Kotani E, Yamakawa M, Iwamoto S, Tashiro M, Mori H, Sumida M, Matsubara F, Taniai K, Kadono-Okuda K, Kato Y, Mori H (1995) Cloning and expression of the gene of hemocytin, an insect humoral lectin which is homologous with the mammalian von Willebrand factor. Biochim Biophys Acta 1260: 245–258

Kubo T, Kawasaki K, Natori S (1993) Transient appearance and localization of a 26-kDa lectin, a novel member of the *Periplaneta* lectin family, in regenerating cockroach leg. Dev Biol 156: 381–390

Lackie AM, Vasta GR (1988) The role of galactosyl-binding lectin in the cellular immune response of the cockroach *Periplaneta americana* (Dictyopetra). Immunology 64: 353–357

Lasky LA (1992) Selectins: interpreters of cell-specific carbohydrate information during inflammation. Science 258: 964–969

Lee RT, Lin P, Lee YC (1984) New synthetic cluster ligands for galactose/N-acetylgalactosamine-specific lectin of mammalian liver. Biochemistry 23: 4255–4261

Lee RT, Rice KG, Rao N BN, Ichikawa Y, Barthel T, Vladimir P, Lee YC (1989) Binding characteristics of N-acetylglucosamine-specific lectin of the isolated chicken hepatocytes: similarities to mammalian hepatic galactose/N-acetylgalactosamine-specific lectin. Biochemistry 28: 8351–8358

Leffler H, Barondes SH (1986) Specificity of binding of three soluble rat lung lectins to substituted and unsubstituted mammalian ß-galactosides. J Biol Chem 261: 10119–10126

Leffler H, Masiarz FR, Barondes SH (1989) Soluble lactose-binding vertebrate lectins: a growing family. Biochemistry 28: 9222–9229

Leprince C, Draves KE, Geahlen RL, Ledbetter JA, Clark EA (1993) CD22 associates with the human surface IgM-B-cell antigen receptor complex. Proc Natl Acad Sci USA 90: 3236–3240

Liao DI, Kapadia G, Ahmed H, Vasta GR, Herzberg O (1994) Structure of S-lectin, a developmentally regulated vertebrate ß-galactoside-binding protein. Proc Natl Acad Sci USA 91: 1428–1432

Linn JJ, Pereira MEA, DeLellis RA, McAdam KPWJ (1984) Human amyloid P component: a circulating lectin that modulates immunological responses. Scand J Immunol 19: 227–236
Liu T-Y, Robey FA, Wang C-M (1982) Structural studies of C-reactive protein. Ann NY Acad Sci 389: 151
Lobsanov YD, Gitt MA, Leffler H, Barondes SH, Rini JM (1993) X-ray crystal structure of the human dimeric S-lac lectin, L-14-II, in complex with lactose at 2.9-Å resolution. J Biol Chem 268: 27034–27038
Loeb JA, Drickamer K (1988) Conformational changes in the chicken receptor for endocytosis of glycoproteins: modulation of binding activity by Ca^{2+} and pH. J Biol Chem 263: 9752–9760
Lotan R (1992) ß-Galactoside-binding vertebrate lectins: synthesis, molecular biology, function. In: Allen HJ, Kisailus C (eds) Glycoconjugates: composition, structure and function. Marcel Dekker, New York, pp 635–671
Lu J, Thiel S, Wiedermann H, Timpl R, Reid KBM (1990) Binding of the pentamer/hexamer forms of a mannan-binding protein to zymosan activates the proenzyme Clr_2 Cls_2 complex of the classical pathway of complement, without involvement of Clq. J Immunol 144: 2287–2294
Ludin C, Hofstetter H, Sargati M, Levy CA, Suter U, Alaimo D, Kilchherr E, Frost H, Delespesse G (1987) Cloning and expression of the cDNA coding for a human lymphocyte IgE receptor. EMBO J 6: 109–114
Marchalonis JJ, Edelman GM (1968) Isolation and characterization of a hemagglutinin from *Limulus polyphemus*. J Mol Biol 32: 453–465
Marschal P, Herrmann J, Leffler H, Barondes SH, Cooper DNW (1992) Sequence and specificity of a soluble lactose-binding lectin from *Xenopus laevis* skin. J Biol Chem 267: 12942–12949
McPhaul M, Berg P (1986) Formation of functional asialoglycoprotein receptor after transfection with cDNAs encoding the receptor proteins. Proc Natl Acad Sci USA 83: 8863–8867
Mecham RP, Hinek A, Griffin GL, Senior RM, Liotta LA (1989) The elastin receptor shows structural and functional similarities to the 67 kDa tumor cell laminin receptor. J Biol Chem 264: 16652–16657
Muchmore AV, Decker JM (1987) Evidence that recombinant IL-la exhibits lectin-like specificity and binds homogeneous uromodulin via N-linked oligosaccharides. J Immunol 138: 2541
Muramoto K, Kamiya H (1990) The amino acid sequences of multiple lectins of the acorn barnacle *Megabalanus rosa* and its homology with animal lectins. Biochim Biophys Acta 1039: 42–51
Muramoto K, Kamiya H (1992) The amino acid sequence of a lectin from conger eel, *Conger myriaster*, skin mucus. Biochim Biophys Acta 1116: 129–136
Muramoto K, Ogata K, Kamiya H (1985) Composition of the multiple agglutinins of the acorn barnacle *Megabalanus rosa*. Agri Biol Chem 49: 85–93
Nguyen NY, Suzuki A, Boykins KA, Liu T-Y (1986a) The amino acid sequence of *Limulus* C-reactive protein. Evidence of polymorphism. J Biol Chem 261: 10456–10465
Nguyen NY, Suzuki A, Cheng S-M, Zon G, Liu T-Y (1986b) Isolation and characterization of *Limulus* C-reactive protein genes. J Biol Chem 261: 10450–10455
Oda Y, Herrmann J, Gitt MA, Turck CW, Burlingame AL, Barondes SH, Leffler H (1993) Soluble lactose-binding lectin from rat intensive with two different carbohydrate-binding domains in the same peptide chain. J Biol Chem 268: 5929–5939
Okino N, Kawabata S, Saito T, Hirata M, Takagi T, Iwanaga S (1995) Purification, characterization, and cDNA cloning of a 27-kDa lectin (L10) from horseshoe crab hemocytes. J Biol Chem 270: 31008–31015
Otter M, Barrett-Bergshoeff MM, Rijken DC (1991) Binding of tissue-type plasminogen activator by the mannose receptor. J Biol Chem 266: 13931–13935
Paroutaud P, Levi G, Teichberg VI, Stromberg AD (1987) Extensive amino acid sequence homologies between animal lectins. Proc Natl Acad Sci USA 84: 6345–6348
Pfeifer K, Hassemann M, Gamulin V, Bretting H, Fahrenholz F, Müller WEG (1993) S-type lectins occur also in vertebrates: high conservation of the carbohydrate recognition domain in the lectin genes from the marine sponge *Geodia cydonium*. Glycobiology 3: 179–184

Pochon S, Graber P, Yeager M, Jansen K, Bernard AR, Aubry J-P, Bonnefoy J-Y (1992) Demonstration of a second ligand for the low affinity receptor for immunoglobulin E (CD23) using recombinant CD23 reconstituted into fluorescent microsomes. J Exp Med 176: 389–397

Powell LD, Varki A (1995) I-type lectins. J Biol Chem 270: 14243–14246

Powell LD, Sgroi D, Sjoberg ER, Stamenkovic I, Varki A (1993) Natural ligands of the B cell adhesion molecule CD22ß carry N-linked oligosaccharides with α-2, 6-linked sialic acids that are required for recognition. J Biol Chem 268: 7019–7027

Raz A, Pazerini G, Carmi P (1989) Identification of the metastasis-associated galactoside-binding lectin as a chimeric gene product with homology to an IgE-binding protein. Cancer Res 49: 3489–3493

Rice KG, Weisz OA, Barthel T, Lee RT, Lee YC (1990) Defined stoichiometry of binding between triantennary glycopeptide and the asialoglycoprotein receptor of rat hepatocytes. J Biol Chem 265: 18429–18434

Richards ML, Katz DH (1990) The binding of IgE to murine Fe RII is calcium-dependent but not inhibited by carboydrate. J Immunol 144: 2638–2646

Roberson MM, Ceri H, Shadle PJ, Barondes SH (1981) Heparin-inhibitable lectins: marked similarities in chicken and rat. J Supramol Struct Cell Biochem 15: 395–402

Robertson MW, Albrandt KA, Keller D, Liu F-T (1990) Human IgE-binding protein: a soluble lectin exhibiting a highly conserved interspecies sequences and differential recognition of IgE glycoforms. Biochemistry 29: 8093–8100

Robey FA, Jones KD, Tanaka T, Liu TY (1984) Binding of C-reactive protein to chromatin and nucleosome core particles. A possible physiological role of C-reactive protein. J Biol Chem 259: 7311–7321

Saito T, Kawabata S, Hirata M, Iwanaga S (1995) A novel type of limulus lectin L6. J Biol Chem 270: 14493–14499

Sato M, Kawakami K, Osawa T, Toyoshima S (1992) Molecular cloning and expression of cDNA encoding a galactose/N-acetylgalactosamine-specific lectin on mouse tumoricidal macrophages. J Biochem 111: 331–336

Schulte RJ, Campbell MA, Fischer WH, Sefton BM (1992) Tyrosine phosphorylation of CD22 during B cell activation. 258: 1001–1004

Scott MT (1971) Recognition of foreignness in invertebrates. II. In vitro studies of cockroach phagocytic haemocytes. Immunology 21: 817–828

Sgroi D, Varki A, Braesch-Andersen S, Stamenkovic I (1993) CD22, A B cell-specific immunoglobulin superfamily member, is a sialic acid-binding lectin. J Biol Chem 268: 7011–7018

Sherblom AP, Decker JM, Muchmore AV (1988) The lectin-like interaction between recombinant tumor necrosis factor and uromodulin. J Biol Chem 263: 5418–5424

Sherblom AP, Sathyamoorthy N, Decker JM, Muchmore AV (1989) IL-2, a lectin with specificity for high mannose glycopeptides. J Immunol 143: 939–944

Shinkai Y, Rathbun G, Lam KP, Oltz EM, Stewart V, Mendelsohn M, Charron J, Stall AM, Alt FW (1992) RAG-2-deficient mice lack mature lymphocytes owing to inability to initiate V(D)J rearrangement. Cell 68: 855–867

Sparrow CP, Leffler H, Barondes SH (1987) Multiple soluble ß-galactoside-binding lectins from human lung. J Biol Chem 262: 7383–7390

Spiess M (1990) The asialoglycoprotein receptor: a model for endocytotic transport receptors. Biochemistry 29: 10008–10019

Stahl PD (1990) The macrophage mannose receptor: current status. Am J Respir Cell Mol Biol 2: 317–318

Stahl PD, Schlesinger PH (1980) Receptor-mediated pinocytosis of mannose/N-acetylglucosamine-terminated glycoproteins and lysosomal enzymes by macrophages. Trends Biochem Sci 5: 1569–1570

Suzuki T, Takagi T, Furukohri T, Kawamura K, Nakauchi M (1990) A calcium-dependent galactose-binding lectin from the tunicate *Polyandrocarpa misakiensis*. J Biol Chem 265: 1274–1281

Takahashi H, Komano H, Kawaguchi N, Kitamura N, Nakanishi S, Natori S (1985) Cloning and sequencing of cDNA of *Sarcophaga peregrina* humoral lectin induced on injury of the body wall. J Biol Chem 260: 12228–12233

Taylor ME, Drickamer K (1993) Structural requirements for high affinity binding of complex ligands by the macrophage mannose receptor. J Biol Chem 268: 399–404

Taylor ME, Conary JT, Lennartz MR, Stahl PD, Drickamer K (1990) Primary structure of the mannose receptor contains multiple motifs resembling carbohydrate-recognition domains. J Biol Chem 265: 12156–12162

Taylor ME, Bezouska K, Drickamer K (1992) Contribution to ligand binding by multiple carbohydrate-recognition domains in the macrophage mannose receptor. J Biol Chem 267: 1719–1726

Tennent GA, Pepys MB (1994) Glycobiology of the pentraxins. Biochem Soc Trans 22: 74–79

Tiemeyer M, Brandley BK, Ishihara M, Swiedler SJ, Greene J, Hoyle GW, Hill RL (1992) The binding specificity of normal and variant rat Kupffer cell (lectin) receptors expressed in COS cells. J Biol Chem 267: 12252–12257

Tyson CJ, Jenkin CR (1974) Phagocytosis of bacteria in vitro by haemocytes from the crayfish *(Parachaeraps bicarinatus)*. Aust J Exp Biol Med Sci 52: 341–348

Uhlenbruck G, Solter J, Janssen E, Haupt H (1982) Anti-galactan anti-haemocyanin specificity of CRP. Ann NY Acad Sci 389: 476–479

Vandenbrule FA, Buicu C, Sobel ME, Liu FT, Castronovo V (1995) Galectin-3, a laminin-binding protein, fails to modulate adhension of human melanoma cells to laminin. Neoplasma 42: 215–219

Vasta GR (1990) Invertebrate lectins, C-reactive and serum amyloid. Structural relationships and evolution. In: Marchalonis JJ, Reinisch CL (eds) Defense molecules. Wiley-Liss, New York, pp 183–199

Vasta GR (1991) The multiple biological roles of invertebrate lectins: their participation in nonself recognition mechanisms. In: Warr GW, Cohen N (eds) Phylogenesis of immune functions. CRC Press, Boca Raton, pp 73–101

Vasta GR (1992) Invertebrate lectins: Distribution, synthesis, molecular biology and function. In: Allen HJ, Kisailus C (eds) Glycoconjugates: composition, structure and function. Marcel Dekker, New York, pp 593–634

Vasta GR, Ahmed H (1995) Lectins from aquatic invertebrates: strategies and methods for their detection, isolation and biochemical characterization – a review. In: Stolen JS, Fletcher TC, Smith SA, Zelikoff JT, Kaattari SL, Anderson RS, Söderhäll K, Weeks-Perkins BA (eds) Techniques in Fish immunology-4, SOS Publications. Fair Haven, pp 241–258

Vasta GR, Marchalonis JJ (1986) Galactosyl-binding lectins from the tunicate *Didemnum candidum*. Carbohydrate specificity and characterization of the combining site. J Biol Chem 261: 9182–9186

Vasta GR, Marchalonis JJ (1987) Invertebrate agglutinins and the evolution of humoral recognition factors. In: Cinader BA (ed) Cell receptors and cell communication in invertebrates. Marcel Dekker, New York, pp 104–117

Vasta GR, Sulivan JT, Cheng TC, Marchalonis JJ, Warr GW (1982) A cell membrane associated lectin of the oyster hemocyte. J Invertebr Pathol 40: 367–377

Vasta GR, Cheng TC, Marchalonis JJ (1984a) A lectin on the hemocyte membrane of the oyster *(Crassostrea virginica)*. Cell Immunol 88: 475–488

Vasta GR, Marchalonis JJ, Kohler H (1984b) An invertebrate recognition protein cross-reacts with an immunoglobulin idiotype. J Exp Med 159: 1270–1276

Vasta GR, Hunt J, Marchalonis JJ, Fish WW (1986a) Galactosyl-binding lectins from the tunicate *Didemnum candidum*. Purification and physiological characterization. J Biol Chem 261: 9174–9181

Vasta GR, Marchalonis JJ, Decker J (1986b) Binding and mitogenic properties of galactosyl specific lectin from the tunicate *Didemnum candidum* for murine thymocytes. J Immunol 137: 3216–3223

Vasta GR, Quesenberry MS, Ahmed H (1996) A lectin from the tunicate *Clavelina picta* is a homologue of the mannose-binding proteins from vertebrates. (In press)

Vavasseur F, Berrada A, Heuze F, Jotereau F, Meflah K (1990) Fucose and galactose receptor and liver recognition by lymphoma cells. Int J Cancer 248: 744–751

Veno PA, Strumski MA, Kinsey WH (1990) Purification and characterization of echinonectin, a carbohydrate-binding protein from sea urchin eggs. Dev Growth Differ 32: 315–319

Vercelli D, Helm B, Marsh P, Padlan E, Geha RS, Gould H (1989) The B-cell binding site on human immunoglobulin E. Nature 338: 649–651

Verrey F, Drickamer K (1993) Determinants of oligomeric structure in the chicken liver glycoprotein receptor. Biochem J 292: 149–155

Watson SR, Imai Y, Fennie C, Geoffroy JS, Rosen SD, Lasky LA (1990) A homing receptor-IgG chimera as a probe for adhesive ligands of lymph node high endothelial venules. J Cell Biol 110: 2221–2229

Wong S, Freeman JD, Kelleher C, Mager D, Takei F (1991) Ly-49 multigene family. New members of a superfamily of type II membrane proteins with lectin-like domains. J Immunol 147: 1417–1423

Woo HJ, Shaw LM, Messier JM, Mercurio AM (1990) The major non-integrin laminin binding protein of macrophages is identical to carbohydrate binding protein 35 (Mac-2). J Biol Chem 265: 7097–7099

Yokoyama W, Jacobs LB, Kanagawa O, Shevach EM, Cohen DI (1989) A murine T lymphocyte antigen belongs to a supergene family of type II integral membrane proteins. J Immunol 143: 1379–1386

Yokoyama WM, Ryan JC, Hunter JJ, Smith HRC, Stark M, Seaman WE (1991) cDNA cloning of mouse NKR-PI and genetic linkage with Ly-49. Identification of a natural killer cell gene complex on mouse chromosome 6. J Immunol 147: 3229–3236

Yuen C-T, Lawson AM, Chai W, Larkin M, Stoll MS, Stuart AC, Sullivan FX, Ahern TJ, Feizi T (1992) Novel sulfated ligands for the cell adhesion molecule E-selectin revealed by the neoglycolipid technology among O-linked oligosaccharides on an ovarian cystadenoma glycoprotein. Biochemistry 31: 9126–9131

Zhou Q, Cummings RD (1990) The S-type lectin from calf heart tissue binds selectively to the carbohydrate chains of laminin. Arch Biochem Biophys 281: 27–35

Ziegler SF, Ramsdell F, Alderson MR (1994) The activation antigen CD69. Stem Cells 12: 456–465

Characterization of the Receptor Protein-Tyrosine Kinase Gene from the Marine Sponge *Geodia cydonium*

W.E.G. Müller and H. Schäcke[1]

1 Introduction

Cells are provided with well-defined receptor structures (signal receivers) which interact with their corresponding ligands (signal molecules) and initiate a signal transduction pathway resulting in a change of cellular behavior or metabolism (Stoddard et al. 1992). It is well established that cells from both eukaryotic protists (single-cell organisms) and from Metazoa (multicellular organisms) respond to signals emanating from the extracellular environment. The extracellular signals to which protists respond are mainly nutrients which diffuse to their surfaces, and in most cases cross the cell membrane. In addition, they are able to bind peptide hormones, e.g., insulin or adrenocorticotropic hormone, as in the unicellular *Tetrahymena*, by receptor-like structures (Köhidai et al. 1994). Based on experimental data obtained with *Tetrahymena*, it has been proposed that the survival of protists presupposes the operation of highly dynamic membrane structures capable of recognizing a variety of environmental signals, interactions which are stored in a form of "memory" and transmitted to the progeny generation (Csaba 1987, 1994). Hence, in unicellular eukaryotes, the membrane-bound receptor(s) have a nondetermined ability to recognize ligands and are initially not genetically programmed. In contrast, in Metazoa the receptors are genetically preprogrammed.

During the transition from the protozoan to the metazoan stage, animals developed genetically encoded cell-surface receptors that interact with a very limited number of ligands (usually only a single one) to ensure a tuned cellular interaction with the environment during the different stages of their development.

Based on nucleotide sequence homology, domain organization, and the proposed signaling mechanisms or function, the cell-surface receptors in Metazoa are usually subdivided into the three main classes: the channel-linked receptor, G-protein-linked receptor, and catalytic cell-surface receptors (Stoddard et al. 1992). Functionally, the receptors are involved in the control of growth and metabolism of a given cell in the tuned interplay with other cells

[1]Abteilung für Angewandte Molekularbiologie, Institut für Physiologische Chemie, Johannes Gutenberg-Universität, Mainz, Duesbergweg 6, 55099 Mainz, Germany

in the organisms. One characteristic feature of Metazoa is the development of an extracellular matrix which is present in all phyla. These structures are already present in the lowest Metazoa, the Porifera (sponges).

Paleontological data suggest that Porifera existed already in the early Cambrian [>600 m.a.] (Knoll 1994). In one view, the Porifera are considered simply as colonies of choanoflagellata which have derived from a separate protist lineage and evolved independently from the deuterostomes, protostomes, and cnidarians (Barnes 1980). In contrast, according to Weissenfels (cited in Mehlhorn 1989), some sponge cells are already organized into epithelia-like tissues which form simple organs or organ-like assemblies. The most prominent examples are the choanocyte chambers consisting of two kinds of epitheliae formed by choanocytes and cone cells. In detail, the cone cells surround the apopyle (beginning of the excurrent channel) and contact with the ap-endopinacocytes and the choanocytes. The apopyle participates in the regulation system of the water current, which is driven by the flagella-beating choanocytes. The choanocyte chambers are embedded in the mesohyl matrix, which is covered by the ap-endopinacocytes and the pros-endopinacocytes. The latter cells contact with the choanocytes at the sites of the prosopyles, which allow the water current to enter the choanocyte chambers. The function of the choanocyte chambers as organs or organ-like assemblies is to orient the water flow into one direction from the incurrent to the excurrent channel systems.

It is well known that sponges are provided with several types of adhesion systems which allow controlled interactions of the cells in the body; the cell-cell and the cell-matrix systems (reviewed by Müller 1982; Müller et al. 1988). However, until recently, no genetic information on cell-surface receptors in these lowest multicellular organisms was available. Here, we describe the structure of the cDNA coding for a receptor protein-tyrosine kinase (RTK), recently identified in the marine sponge *Geodia cydonium*. This sponge belongs to the subphylum Cellularia (class: Demospongiae). The sponge RTK shows the characteristic features of homologous molecules known from higher animals. A nonreceptor tyrosine kinase, the *src*-related tyrosine kinase gene, has been analyzed from the freshwater sponge *Spongilla lacustris* (Ottilie et al. 1992).

2 Protein Kinases

Protein kinases are enzymes that transfer a phosphate group from a phosphate donor (the γ phosphate of a nucleoside triphosphate) onto an amino acid (aa) in a substrate protein (Hunter 1991). The different kinases are classified according to the acceptor aa for the phosphate group: (1) Serine- and threonine-specific protein kinases (generating phosphate esters at the alcohol group of the respective aa); (2) tyrosine-specific protein kinases (generating phosphate esters with the phenolic group of tyrosine); (3) protein-histidine kinases (generating phosphoramidates with histidine, arginine, or lysine); (4) protein-cysteine kinases (generating phosphate thioesters), and (5) protein-aspartyl/glutamyl kinases (generating mixed phosphate-carboxylate acid anhydrides).

The first two classes of kinases have been well characterized; both the serine/threonine-specific and the tyrosine-specific protein kinases occur in pro- and eukaryotes (reviewed by Hanks and Quinn 1991). The boundaries of the catalytic domains of the serine/threonine-specific and the tyrosine-specific protein kinases are defined at the level of deduced aa sequences which are highly conserved. The amino-terminus, the boundary of both classes of kinases, starts seven aa residues upstream from the first Gly in the consensus Gly-Xxx-Gly-Xxx-Xxx-Gly with a cluster of hydrophobic residues; the carboxyl-terminus is set at a hydrophobic aa residue, positioned ≈10–17 residues downstream of the invariant Arg. The segment between the two boundaries spans about 550 aa residues.

The catalytic domain of the tyrosine-specific protein kinases has been further divided – according to regions of localized high conservation of aa residues – into 12 subdomains (Hanks and Quinn 1991) which are separated by regions of low conservation. As outlined, such an arrangement of alternating regions of high and low conservation is typical for homologous globular proteins (Hanks et al. 1988). In addition, it has been suggested that the conserved subdomains are involved in the catalytic function – either directly as a part of the catalytic site or indirectly by contributing to the formation of the secondary structure – while the nonconserved regions might comprise the loops allowing the conserved subdomains to interact.

The serine/threonine-specific protein kinases are distinguished from the tyrosine-specific kinases by consensus sequences; the serine/threonine-specific enzyme shows in subdomain VIII the consensus Gly-Thr/Ser-Xxx-Xxx-Tyr/Phe-Xxx-Ala-Pro-Glu, while the tyrosine-specific kinases have the consensus Pro–Ile/Val-Lys/Arg-Trp-Thr/Met-Ala-Pro-Glu at the same region. In addition, the catalytic domains of the kinases harbor the ATP-binding site which will be mentioned in Section 4.5 (Hanks et al. 1988). The exact positions of the aa within the sponge sequence are given in Section 4.5.

3 Receptor Protein-Tyrosine Kinases

Receptor protein-tyrosine kinases [RTKs] have been described in higher as well as in lower invertebrate species (e.g., fish, electric ray, and amphibians; reviewed in Geer et al. 1994). RTKs are present also in some invertebrates. Many of these have been isolated from *Drosophila*; among them are the epidermal growth factor receptor-related DER (Livneh et al. 1985), the insulin receptor-related DILR (Nishida et al. 1986), and the fibroblast growth factor receptor-related DFR1 (Shishido et al. 1993) and also from *Caenorhabiditis elegans*, e.g., *let*-23 (Aroian et al. 1993) and *kin*-15 and *kin*-16 (Morgan and

Greenwald 1993). Furthermore, putative RTK genes have been found in the coelenterate *Hydra* (Geer et al. 1994).

Recently, we described a RTK gene from the siliceous sponge *Geodia cydonium* (Schäcke et al. 1994a). The deduced aa sequence shows all the characteristic features known from vertebrate sequences.

The RTKs belong to group 1 receptors and have a single transmembrane segment with their N-termini outside the cells. These receptors have distinct segments starting with signal peptide – the extracellular domain (ligand-binding domain) – followed by the transmembrane domain – the juxtamembrane domain – and finally the catalytic domain.

To the present, the following 14 groups of RTKs have been identified and classified according to their external domain; they have been termed subfamilies (reviewed by Hunter et al. 1992; Geer et al. 1994) or classes (Ullrich and Schlessinger 1990). The subfamily I, epidermal growth factor receptor (EGFR; class I), contains two Cys-rich regions; the members of subfamily II, platelet-derived growth factor receptors (PDGFR; class III), have five or seven immunoglobulin-like (Ig-like) domains. The subfamily III, insulin receptor (INSR; class II) has, in addition to the Cys-rich domain, three fibronectin (FN) repeats; it is derived from a precursor molecule from which the α and β-subunits are formed by proteolytic cleavage, and forms heterotetrameric molecules. The nerve growth factor receptor (NGFR), subfamily IV, has two Ig-like domains and in addition a Leu-rich motif; members of subfamily V, fibroblast growth factor receptor (FGFR; class IV), exhibit three related sequence repeats which show weak homology to the interleukin 1 receptor, a member of the Ig superfamily. The EPH, subfamily VI, has one Ig-like domain, a distinct Cys-rich region, and three FN repeats, while the oncogenic UFO (AXL), subfamily VII, has two FN repeats and two Ig-like domains, and TIE, subfamily VIII, contains the Ig-like and FN-domains and three EGF repeats. A further domain, characteristic for cadherin, is found in RET, subfamily IX. The hepatocyte growth factor receptor (HGFR), subfamily X, is a heterodimer which is formed from a precursor by proteolytic cleavage and linked together by disulfide bonds. Kgl, subfamily XI, has seven Ig-like domains and ROR1, subfamily XII, has, in addition to a kringle motif and the Ig-like domain, also a unique Cys-rich region. The oncogenic RYK (EYK) subfamily XIII, has two Ig-like and two FN-domains, and the DDR, subfamily XIV, is provided with a discoidin I-like domain.

Published data show that only short aa sequences in the extracellular part of the RTKs interact with the corresponding ligands, suggesting that the mentioned domains might have additional functions, e.g., in dimerization of two receptor molecules one prerequisite for RTK signaling, in interaction with other cell surface molecules to promote adhesion, communication between cells, or binding to the extracellular matrix (Geer et al. 1994).

The intracellular catalytic domain – tyrosine kinase – is highly conserved and displays its enzymic function after ligand binding. Two kinds of tyrosine phosphorylation occur, autophosphorylation and phosphorylation of cytoplasmic substrates. These reactions are reversible; dephosphorylations of the tyrosyl residues are catalyzed by protein tyrosine phosphatases (PTPs). Tyrosine kinases and tyrosine phosphatases operate in an antagonistic or co-operative way during signal transduction (reviewed in Sun and Tonks 1994).

Ligand binding is assumed to cause activation of RTK via either stabilization or induction of receptor dimerization. As a result, the two catalytic domains of the receptors are juxtaposed, allowing intermolecular autophosphorylation; most of the autophosphorylation sites lie outside the catalytic domain. The autophosphorylation sites of RTKs are probably involved in the selection of proteins that contain Src-homology 2 (SH2) domains (Hunter et al. 1992; Schlessinger and Ullrich 1992; Geer et al. 1994; Sun and Tonks 1994). This binding is specific; especially crucial are the (three to five) aa C-terminal to the P•Tyr; these aa interact with the SH2 domain.

The substrates for the RTKs have been classified into three classes; (1) enzymes which undergo alterations in their enzymatic function in dependence on phosphorylation [phospholipases C, A_2 and D; phosphatidylinositol 3-kinase; protein tyrosine phosphatases; c-*src* family members], (2) proteins without a catalytic activity, e.g., SH2 [the oncogenic proteins CRK, NCK and SHC; and the adapter molecule GRB2] or SH3 and (3) other RTK docking proteins such as insulin receptor substrate-1 [IRS-1] (Glenney 1992; Geer et al. 1994).

4 Receptor Protein-Tyrosine Kinase from the Sponge *Geodia cydonium*

Screening of the cDNA library from the sponge *G. cydonium* was performed under low stringency hybridization conditions (Schäcke et al. 1994a) using the human insulin-like growth factor I receptor sequence (Ullrich et al. 1986). Two species of RTKs were identified. In the present chapter the sequences of the two representative clones GCTK_1 and GCTK_2 are described in detail. The total length of GCTK_1 is 2066 nt and of GCTK_2 2470 nt. These sizes correspond to 639 aa (GCTK_1) and 676 aa (GCTK_2), respectively. The distribution of the different classes of aa are for GCTK_1 (GCTK_2): acidic 11.4% (10.4%), basic 17.1% (11.0%), aromatic 7.7% (7.7%), and hydrophobic 33.4% (32.9%); a scheme of the sequences is given in Fig. 1.

The size of the sponge RTK mRNA was identified by Northern blotting using the cDNA of GCTK_2 as a probe. Only one mRNA species of 3.3 kb was detected (Fig. 2). This size of the transcript indicates that the RTK sequences cloned from *G. cydonium* are not yet complete. The start methionine is also missing.

The RTK from *G. cydonium* represents a novel class of receptors because the catalytic domain is very similar to the subfamily III (according to Hunter et al. 1992) or class II (according to Ullrich and Schlessinger 1990) RTKs which have no Ig-like domain. The sponge RTK contains both the class II

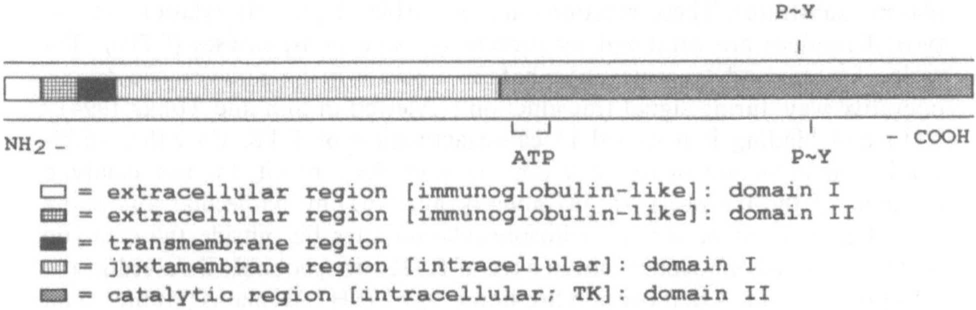

Fig. 1. Structure of receptor protein-tyrosine kinase gene from the marine sponge *Geodia cydonium*. The gene is divided into five blocks; the two extracellular domains (comprising two Ig-like domains), the transmembrane region, the juxtamembrane region (intracellular domain I), and the catalytic region [intracellular TK (tyrosine kinase) domain]. The relative positions of the ATP-binding site and the phosphorylation, as well as the autophosphorylation sites, are given

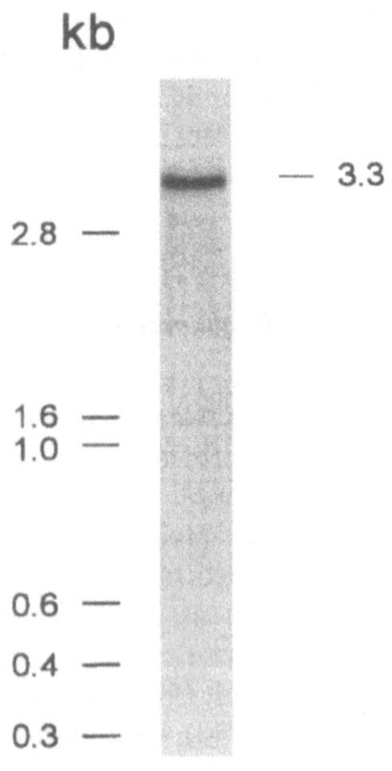

Fig. 2. Identification of *G. cydonium* RTK mRNA by Northern blotting. RNA was isolated and 10 μg of RNA was resolved by electrophoresis in 1.2% agarose/0.66 M formaldehyde gel. The RNA was transferred to Nylon 66 and hybridized with the GCTK_2 clone [TK catalytic domain (the 432 bp Pst I - Sal I fragment)]. The filters were exposed to Kodak X-Omat film

catalytic- and the Ig-like domain. The cDNA is arranged blockwise; therefore the different parts of the sponge RTK are discussed separately.

Recent data obtained from screening of a genomic library are included (Skorokhod, pers. comm.) to obtain first information about intron sequences in sponges and results about the arrangement of the Ig-like domains.

4.1 Ligand-Binding Domain (Immunoglobulin-Like Domain)

The deduced aa sequences of the extracellular part of the sponge RTK comprises two Ig-like domains. They are present in complete length in the clone GCTK_2; only one domain was covered by GCTK_1, while the exon/intron sequence covers both domains (Fig. 3). Sequence comparison revealed that both GCTK_1 and the exon sequence differ from GCTK_2 on the nt and the aa level, suggesting three genes.

The most conserved region is the immunoglobulin fold which is composed of approximately 110 aa residues and is arranged into two parallel β-sheets which are stabilized by one disulfide bond. The Ig domains are related and are classified on the basis of their folding patterns into the V (variable) and C (constant) domains (Williams and Barclay 1988). The sequences of the Ig domains are grouped into the categories V-SET, C1-SET and C2-SET. The core of the Ig fold for the domains consists of β-strands A, B, E, and D in one sheet and G, F, and C in the other. The two β-sheets forming the sandwich of the Ig structure are stabilized by a conserved disulfide bond spanning from β-strand B to β-strand F. The consensus pattern for Ig chains is (FY)-x_1-C-x_2-(VA)-x_3-x_4 around the C-terminal Cys; the amino acid representing x_4 in the V-SET is not conserved, in the C1-SET it is His and in the C2-SET it is a "small" amino acid.

Figure 4 shows that the highly conserved amino acid Cys in the β-strand B is present in all sequences; in addition, the conserved Trp in the β-strand C exists 12–14 amino acids apart from Cys. Adjacent to Cys are the characteristic amino acids Leu and Ile. The second characteristic pattern is in the β-strand F. Here, the consensus pattern in the sponge RTK is D-x-G-x-Y-x-C, which matches perfectly with the vertebrate and insect sequences. Adjacent to Cys are one hydrophobic amino acid (Val in the sponge) and one small amino acid (Gly in the sponge). The results show that, already in sponges, an Ig-like domain with a β-strand of the C2-like type is found in the receptor molecule (Schäcke et al. 1994b, c).

The sponge sequence (second Ig-like domain) was compared both with the following invertebrate and vertebrate second Ig-like domains (Fig. 4); hemolin I (soluble hemolymph protein from the moth *Hyalophora cecropia*; Sun et al. 1990), neuroglian I (a homophilic adhesion molecule which mediates the adhesion between neurons and glia; Seeger et al. 1988), fasiclin II (a homophilic adhesion molecule involved in pathway recognition for axons) both from *Drosophila melanogaster* (Harrelson and Goodman 1988), amalgam (the cell adhesion molecule of the Antennapedia complex; Seeger et al. 1988) as well as the selected vertebrate sequences chicken N-cell adhesion molecule (homophilic binding molecule which mediates the interactions between neurons and other brain cells), human fibroblast growth factor receptor BFR-2 (a receptor supporting the growth of endothelial cells), human keratinocyte growth factor receptor (a receptor present at epithelial cells which is ligand-activated), as well as human alpha platelet-derived growth factor receptor (a receptor at con-

```
GCTK_2  GACCTGAGCCAGCCACACTCTGTCACTCTCACTTGCTCTGCCACCAGTCCTCCTGCCCGTGCGTACCAATACCAGTGGCAGTGGAGGAGG  90
         D  L  S  Q  P  H  S  V  T  L  T  C  S  A  T  S  P  P  A  R  A  Y  Q  Y  Q  W  Q  W  R  R   30

        [- - - I - - -
GCTK_2  AATGGGACACTACTGAGCAACACATACACTAGATTCTCTATCACACCCTCCACCAACACTCAGTCCAGTAGTCTAGTCATATCTGGTCTC 180
         N  G  T  L  L  S  N  T  Y  T  R  F  S  I  T  P  S  T  N  T  Q  S  S  S  L  V  I  S  G  L   60
EXON    ••••••••••••••••••••••••C••••••••••••••••••T••••••••••••••••••••••••••••••••••••••••••••••
         •  •  •  •  •  •  •  •  H  •  •  •  •  •  I  •  •  •  •  •  •  •  •  •  •  •  •  •  •  •

                             † intron
GCTK_2  AGATATCCTGATGCAGGAGACTACATGTGTACAGTGAAGTATGGACCATGTCCTGATGGAGGTGACTGCGGTGGAACAACTCCAGTCACT 270
         R  Y  P  D  A  G  D  Y  M  C  T  V  K  Y  G  P  C  P  D  G  V  D  C  G  G  T  T  P  V  T   90
EXON    ••G•••T•••••••••••••••••••••••••••••G••••••••GT•••••••••••••••T••••••A••••••••••••••••••••
         •  •  S  •  •  •  •  •  •  •  •  •  E  •  •  V  •  •  •  •  •  •  •  S  •  •  •  •  •  •

GCTK_2  GGAGTCATACATCTTGAACTTCCATTGATAGTGGAGGTGGATTCCTCTGGGCTGGTGGTGAGAGAAGGAAGTGAGGTGATTGTTCTGACA 360
         G  V  I  H  L  E  L  P  L  I  V  E  V  D  S  S  G  L  V  V  R  E  G  S  E  V  I  V  L  T  120
EXON    ••••••T••••••••••C••C••••••••••••••••••••C••••••••••••A•••••••••••••••••••••••••••••••••••
         •  •  L  •  •  D  •  •  •  •  •  •  •  •  •  •  •  •  M  •  •  •  •  •  •  •  •  •  •  •

                                 † intron
        - - - - I - - -] [- - - II - - -
GCTK_1                       •••••••••••••••••••••••••••••••••••••••••••••••••••••••••••••••••••••
                                  •  •  •  •  •  •  •  •  •  •  •  •  •  •  •  •  •  •  •  •  •  •  •

GCTK_2  TGTGAGGTGTACGGCTATCCTCGAGACTCCTCCCCTCCCATGTGGAGCTCTCCTGGGAGAAACCTGGAGTCTGGCAGATTCAACATTACT 450
         C  E  V  Y  G  Y  P  R  D  S  S  P  P  M  W  S  S  P  G  R  N  L  E  S  G  R  F  N  I  T  150
EXON    ••••••••••••••T•••••••••••••••••T••A•••••••••••••••••••••••••••••••••••••••••••••••••C••••
         •  •  •  •  •  •  •  •  •  •  •  •  •  •  •  •  •  •  •  •  •  •  •  •  •  •  •  •  T  •
```

```
GCTK_1  •••••••••••••••••••••••••••••••••••••••••••••••••••••••••••••••••••••••••••••••••GTT••••••
         •  •  •  •  •  •  •  •  •  •  •  •  •  •  •  •  •  •  •  •  •  •  •  •  •  •  •  V  •  •
GCTK_2  CCCAGATACACTGGCACACTGAGCAATGGCAGTGTGTCCTCCTCTGATAAAGTGGCCCTGTCTCAACTCACCATATTCAACATCACTGTG 540
         P  R  Y  T  G  T  L  S  N  G  S  V  S  S  S  D  K  V  A  L  S  Q  L  T  I  F  N  I  T  V  180
EXON    •••••A••••••••••••••••••••••••••••••••••••••••••••C••G
         •  K  •  •  •  •  •  •  •  •  •  •  •  •  •  •  •  •

GCTK_1  •••••••••••••••••••CA••••••••••••••••••••••••••••••••••••••••••••••A••••••••••••••••••••••
         •  •  •  •  •  •  T  •  •  •  •  •  •  •  •  •  •  •  •  •  •  •  D  •  •  •  •  •  •  •
GCTK_2  GCTGATGAAGGAGAGTACAAGTGTTCAGTGGATGGGGAATCTGCTTCCTTCCGTGTTGATCTAGGTGGCTCCAACTCAAGTGGCAGTAAC 630
         A  D  E  G  E  Y  K  C  S  V  D  G  E  S  A  S  F  R  V  D  L  G  G  S  N  S  S  G  S  N  210

GCTK_1  •••••••A•••••••••••
         •  •  E  •  •  •
GCTK_2  AGCGGAGTGATTGCTGGA 649
         S  G  V  I  A  G  216
         -  -  II -  -  -]
```

Fig. 3. Ligand-binding region (Ig-like domain) of sponge RTK. The two Ig-like domains from GCTK_2, which are marked by *I* and *II*, are compared with the Ig-like domain of GCTK_1 as well as with the corresponding intron/exon sequence obtained from cloning this segment from a genomic library [*EXON*]. Identical nt as well as aa are marked by *filled circles*; gaps are indicated by *hyphens*. The two splice sites are marked †

ß-Strand	B	C	D	E	F
	□□□□□□□□□	□□□□□□□□□	□□□□□□	□□□□□□□□□	□□□□□□□□□
GCTK_2 (1)	*TLTCSATSP*	*AYQYQWQWRRNGTLLSNTY*	*FSITPSTNTQSS*	*VISGLRYP*	*DAGDYMCTVKY*
GCTK_2 (2)	*VLTCEVYGY*	*RDSSPPMWSSPGRNLESGR*	*FNITPRYTGTLS*	*DKVALSQLTIF*	*ITVADEGEYKCSVDG*
Invertebrates:					
Hemolin I	**VLECIIEGN**	**DQGVKYSWKKDGKSYNWQEH**	**NAALRKDE**	**GSLVFLRP**	**QASDEGHYQCFAET**
Neuroglian I	**IIECEADGQ**	**PEPEYSWIKNGIIFDWQ**	**AYDNRMLRQP**	**GRGTLVITIP**	**KDEDRGHYQCFASN**
Fascicl II(2)	**KIRCRVSANP**	**PAIVNWMRDGHIVETG**	**DRYVV**	**EQDGLTIILNVTE**	**MDDGTYTCRAIV**
Amalgam	**ELECSVQGYP**	**APTVVWHKNGVPL**	**QSSRHHEVA**	**ASSSGTTTSVLR**	**VGEEDFGDYYCNATN**
Vertebrates:					
N-CAM V	**NITCEVFAY**	**PSATISWFRDGQLLPSS**	**NYSNIKIYNTPS**	**ASYLEVTPD**	**SENDFGNYNCTAVN**
H-FGFR	**EFVCKVYSDA**	**QPHIQWIKHVEKNGSKY**	**GLPYLKVLKHS**	**NAEVLALFNV**	**TEADAGEYICKVSN**
H-KGFR	**EFVCKVYSDA**	**QPHIQWIKHVEKNGSKY**	**GLPYLKVLKHS**	**NAEVLALFNVT**	**EADAGEYICKVSN**
H-PDGFR	**VVTCAVFNNEV**	**VDLQWTYPGEVKGKGIT**	**MLEEIKVPSIK**	**LVYTLKVPEATV**	**KDSGGYECAAR**
consensus	*	*	*	*	* * * *
			S - S		

Fig. 4. Aligned amino acid sequences of the extracellular domain of RTK sequence from the sponge *Geodia cydonium* [clone GCTK_2 (segments *1* and *2*)] and immunoglobulin-like domains of the following (*i*) invertebrate sequences containing Ig-like domains: hemolin I, neuroglian I, fasciclin II [Fascicl II(2)], and amalgam, as well as (*ii*) the vertebrate species: chicken N-cell adhesion molecule (N-CAM V; Sun et al. 1990), human fibroblast growth factor receptor BFR-2 (H-FGFR), human keratinocyte growth factor receptor (H-KGFR), human alpha platelet-derived growth factor receptor (H-PDGFR; Schäcke et al. 1994b). The highly conserved regions in ß-strands B/C and F (*squares*) and the respective conserved aa are indicated. The consensus marks the identical or related aa characteristic for Ig-like sequences. The disulfide bond is indicated

nective tissue and smooth muscle cells, binding a set of growth factors). The high homologies, especially around the two Cys from which the disulfide bridge originates, are evident (Fig. 4).

It remains open if the sponge RTK represents an adhesion molecule interacting either in homo- or heterophilic fashion or a growth factor which binds to a hitherto nonidentified ligand.

4.2 Intron/Exon

The intron/exon parts of the Ig-like domain of the sponge RTK were cloned to elucidate if the multidomain Ig-like molecule is coded by individual exons, as known for some related molecules from higher organisms; some exceptions are known, e.g., the N-CAM from chicken (Cunningham et al. 1987). The comparison of the deduced aa sequence of GCTK_2 with the exon/intron segment of the same region shows that the splice site is present within the two Ig-like domains (Fig. 3).

One complete exon with flanking introns belonging to exon 1, 2, and 3 was sequenced (Fig. 5). The complete exon/intron segment of RTK (termed EXON-2) shows the typical conserved ends of the intron with GT at the 5′ site and AG at the 3′ terminus (Fig. 5). However, the exon/intron crossing A*T*/GT (C*C*/GT) and the intron/exon crossing AG/*T*G (AG/*A*A) are unusal with respect to the exon nt if compared with the human splice sites A*C*/GT and AG/*G*N (Stephens and Schneider 1992). The intron comprises one inverted repeat and one palindromic sequence. In addition, the intron shows homology (≈50%) with *Dictyostelium discoideum* DRE1 mobile genetic [retrotransposable] element found to be associated at its 3′ ends with tRNA genes. Core element of this *D. discoideum* transposable element is CGCAAAATTCTCAAAA (Marschalek et al. 1993), which matches – with two exchanges – the sponge sequence CGCAAAATT-CACAAAA. The sponge intron shows a pronounced secondary structure with a calculated free energy of structure Δ G of −52.8 kcal.

One potential branch site is present which comprises the sequence CCTCTAC and matches – with one exception – the consensus for higher eukaryotes which is Py_{80}-N-Py_{80}-Py_{87}-Pu_{75}-A-Py_{95}; this sequence precedes the acceptor site 16–10 nt upstream (Fig. 5). Between the potential branch site and the acceptor site a polypyrimidine [oligo$(C)_5$] is inserted.

4.3 Transmembrane Domain

The putative transmembrane regions of the deduced aa sequences of the two clones is 84 (GCTK_1) (81; GCTK_2) nt long and differs by one aa from each other; 87% of the aa are hydrophobic in nature (Fig. 6). They are located 712 nt upstream from the TK catalytic domain. Adjacent to this domain is the Arg-Tyr-Arg-Arg-Arg-Gly-Lys (RYRRRGK) sequence with the three basic aa RRR indicative of a carboxy-terminal location on the cytoplasmic side of the membrane (Ullrich et al. 1985; Van Heijne 1987).

```
        ValSerSerSerAspLysValAla
Exon-3  GTGTCCTCCTCTGACAAGGTGGCCCgtctcaactcaccat
        GlyValLeuHisLeuAspLeuProL    donor site
EXON-2  GGAGTCTTACATCTTGACCTCCCATgtacgtgttgtgtgctcagggtataccgtatagtgcttatatttcccggatctaaattttcgcaa
EXON-2  atagctgttttgaaggaattcgttgaaaaaaattcacagatgcgtgcgtgcagcgcaggggctcaaattttagctgaattaatttcgcaa
EXON-2  aattcacaaaattaaagatataagcactatatgctggcagtgagcgtatttgtgtccactaaacttacggcagaaaccaaaacttctgct
EXON-2  tctgtttcctatctgtgtaaattggtgaataggtgacccccctgtgtgctgctgaaaagcctccaatagcatgtggtgctcaatgtatgga
EXON-2  aaatgggttttccacctaacaggaaaccgtaaactttagtgtagggccaaataagcttatttagattgtgtattatgcgctttcactgtt
EXON-2  ttgctgtactttgtctccctctacccccctatagTGATAGTGGAGGTGGACTCCTCTGGG
                          acceptor site    euIleValGluValAspSerSerGly
                                           AsnThrHisThr
EXON-1  gaggaatggacactactacaacacacacactagAACACACACACT
```

Fig. 5. The exon (*capital* nt letters)-intron region (*small* nt letters) present within the extracellular domain of RTK of *G. cydonium* Exon-2 (*EXON- 2*) was completely sequenced. The flanking segments intron/exon-1 and intron/exon-2 are marked (*EXON-1* and-*2*) The palindromic sequence is *underlined* and marked in *italics*, while the inverted repeat is *double underlined*. The retrotransposable-like element is in *superscript* and is *double underlined*. The donor as well as acceptor sites are *boxed*. *Three-letter code* for the aa is used (A. Skorokhod, pers comm)

```
GCTK_1  GTTTTAATAACTCTTCTATTACTCATAGCCCTCATCATAATTCTCATATGCGTCTTCTGGGTGGTGTGGCGCTACCGCAGAAGA  84
GCTK_2  •••••••••••••••T•••••••••••••••••••••••••••••••••T••••••••••••••••---T••••••••••••••  81
GCTK_1  V  L  I  T  L  L  L  L  I  A  L  I  I  I  L  I  C  V  F  W  V  V  W  R  Y  R  R  R  28
GCTK_2  •  •  •  •  •  •  •  •  •  •  •  •  •  •  •  •  F  •  •  •  •  •  -  C  •  •  •  •  27
```

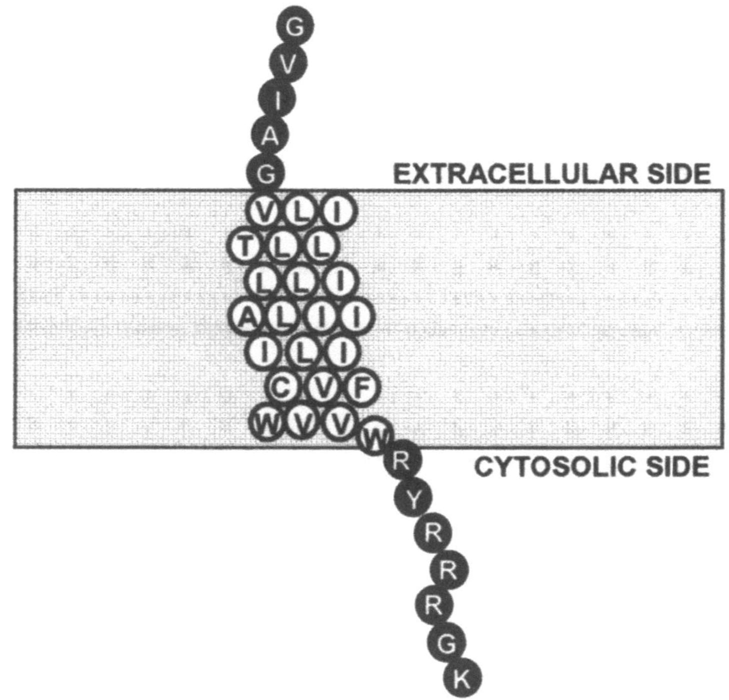

Fig. 6. Transmembrane domain of sponge RTK. The domain is flanked at its carboxy terminus by a stretch of charged/polar amino acids that may facilitate plasma membrane anchoring (*double underlined*). The schematic insertion of the molecule into the cell membrane is also given

```
GCTK_1  GGGAAGTTTGACTTGGGGTCTTGCAGGGAGCTCTCCTGTAGCTCATGTTCGTGTGTGCCTCTCCTTGCTGCACTGAAAGGTGTCAAACTC   90
GCTK_2  •••••••G••••••••A•••••••••••••••••••G•••••••••••••••••••••••••••••••••••••••••••••••••••••   90
GCTK_1   G  K  F  D  L  G  S  C  R  E  L  S  C  S  S  C  S  C  V  P  L  L  A  A  L  K  G  V  K  L    30
GCTK_2   •  •  L  •  •  •  •  •  •  •  •  •  •  G  •  •  •  •  •  •  •  •  •  •  •  •  •  •  •  •    30

GCTK_1  CCAACAAGACACCGAGAAAACTTGAACAAGAACGGAACAAGACTGAGACTGAACGAGAGGAATCATATCGCAGACACCAATACTGAGATT  180
GCTK_2  ••••••••••••••••••••••••G••••••••••••••••••••••••••••••••••A•••••C••••••••••••••••••••••••  180
GCTK_1   P  T  R  H  R  E  N  L  N  K  N  G  T  R  L  R  L  N  E  R  N  H  I  A  D  T  N  T  E  I    60
GCTK_2   •  •  •  •  •  •  •  •  D  •  •  •  •  •  •  •  •  •  •  •  •  •  •  •  •  •  •  •  •  •    60

GCTK_1  TACAGTGTCGTACAGAAACCACTCAAGAAAATCAGCAAATCCCCACCACCACTTCCTCCCCTCACACTCACAGAGACTGAGTTGAATGAA  270
GCTK_2  ••••••••••••••••••••••••••••••••••••••••••••••••••••••••••••••••••••••••••••••••••••••••••  270
GCTK_1   Y  S  V  V  Q  K  P  L  K  K  I  S  K  S  P  P  P  L  P  P  L  T  L  T  E  T  E  L  N  E    90
GCTK_2   •  •  •  •  •  •  •  •  •  •  •  •  •  •  •  •  •  •  •  •  •  •  •  •  •  •  •  •  •  •    90

GCTK_1  CTAATGAGCATAGATGAAAAGGAAGAACTGTCCCCCATCCAAGAGAAACCGACACGAAGAAACACTGGTCTCTCGAC-TACTCTCAGTCA  359
GCTK_2  ••••C••••••••••••••••••••••••••••••••••••••••••••••••••••••••••••••••••••••••C••••••••••••  360
GCTK_1   L  M  S  I  D  E  K  E  E  L  S  P  I  Q  E  K  P  T  R  R  N  T  G  L  S  T  T  L  S  Q   120
GCTK_2   •  T  •  •  •  •  •  •  •  •  •  •  •  •  •  •  •  •  •  •  •  •  •  •  •  •  Y  S  Q  S   119
                                                                                     †
```

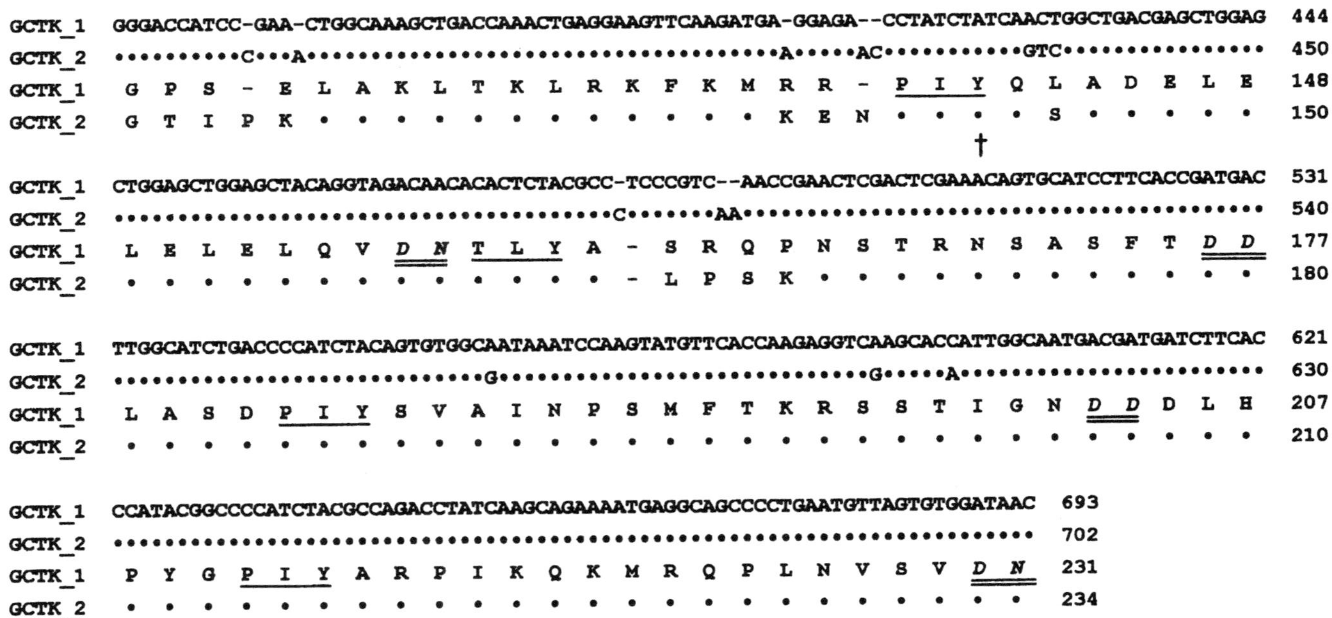

Fig. 7. Juxtamembrane region (cytoplasmic domain I) of sponge RTK. Identical nt as well as aa are marked by *filled circles*; gaps are indicated by *hyphens*. The potential tyrosine phosphorylation sites are marked by *crosses* †. The four apparent repeats are marked: start is *underlined*, and the terminus is in *italics* and *double underlined*

4.4 Juxtamembrane Region

Tyrosine kinase phosphorylation sites are generally characterized by an Arg or Lys positioned seven aa residues to the N-terminal side of the phosphorylated Tyr (Cooper et al. 1984). Two of them are also present in the juxtamembrane region of the sponge RTK; at Tyr^{116} and at Tyr^{143} (Fig. 7). No insulin receptor-specific kinase motif, Tyr-Met-Xxx-Met (Shoelson et al. 1992), is found in the sponge deduced aa sequence of RTK.

The juxtamembrane region shows four apparent repeats which start with the aa trimers Pro(Thr)-Ile(Leu)-Tyr [P(T)-I(L)-Y]; Pro and Thr, as well as Ile and Leu, are related to aa (Fig. 8). Strikingly homologous are also the C-termini of these four segments with Asp-Asp(Asn) [D-D(N)].

4.5 Catalytic Domain

The deduced aa sequence of sponge RTK from the clones of both GCTK_1 and GCTK_2 shows the characteristic features of a protein-tyrosine kinase (Figs. 9, 10) with one exception. Subdomain X, which is in other sequences not very conserved (Hanks et al. 1988), is absent in the sponge catalytic domain, only 11 out of 12 are found. All aa present in serine/threonine- and tyrosine-specific protein kinases described (Hanks et al. 1988) are found also in the sponge sequence at the same positions and in the identical subdomains (Fig. 10); Gly^{10}, Lys^{48}, Glu^{65}, Asp^{151}, Asn^{156}, Asp^{169}, Gly^{171}, Glu^{198}, and Arg^{274}; also those found in almost all kinases exist: Gly^{8}, Val^{159}, Phe^{170}, Asp^{209}, and Gly^{214}. It possesses the core motif of the TK signature Asp-Leu-Ala-Thr-Arg-Asn [D-L-A-T-R-N, at aa no. 151–156 (GCTK_1); subdomain VIb] which is almost identical with the typical TK signature Asp-Leu-Ala-Ala-Arg-Asn (D-L-A-A-R-N, characteristic for both vertebrate and invertebrate TKs; Hanks et al. 1988; Ottilie et al. 1992) Ala and Thr are both "small" aa. This signature differs from the consensus Asp-Leu-Lys-Pro-Glu-Asn (D-L-K-P-E-N) typical for serine/threonine protein kinases (Hanks et al. 1988).

GCTK_1	PIYQLAD----ELELELELQVDN	157
GCTK_2	PIYQSAD----ELELELELQVDN	159
GCTK_1	TLYA-SRQP--NSTRNSASFTDD	177
GCTK_2	TLYALPSKP--NSTRNSASFTDD	180
GCTK_1	PIYSVAINPSMFTKRSSTIGNDD	204
GCTK_2	PIYSVAINPSMFTKRSSTIGNDD	207
GCTK_1	PIYARPIKQKMRQPLNVS--VDN	231
GCTK_2	PIYARPIKQKMRQPLNVS--VDN	234

Fig. 8. Apparent aa repeats within the juxtamembrane region. The alignment starts with identical or related aa trimers; the C-terminus shows also homologies. Gaps are indicated by *hyphens*

Additionally, the deduced aa sequence of the sponge TK contains in subdomain VIII the tyrosine kinase catalytic domain indicator Pro-Ile-Arg-Trp-Met-Ala-Thr-Glu (PIRWMATE; aa no. 191–198) almost identical with the consensus for TKs, Pro-Ile/Val-Lys/Arg-Trp-Thr/Met-Ala-Pro-Glu [P-(IV)-(KR)-W-(TM)-A-P-E]; Thr and Pro are "small" aa. This consensus differs from the corresponding sequence Gly-Thr/Ser-Xxx-Xxx-Tyr/Phe-Xxx-Ala-Pro-Glu [G-T(S)-x-x-Y(F)-x-A-P-E] found in serine/threonine protein kinases (Hanks et al. 1988). Ala-Pro-Glu in the mentioned domain indicator is often referred to as key catalytic site (Hanks et al. 1988).

The protein encoded by sponge RTK bears also hallmarks of other TKs, the ATP-binding site [consensus: Gly-Xxx-Gly-Xxx-Gly (G-X-G-X-X-G); in *G. cydonium*: Gly-Val-Gly-Gln-Phe-Gly (G-V-G-Q-F-G), aa 8–13; subdomain I], and the Lys (residue aa 48; subdomain II) in the consensus Val-Ala-Val-Lys (V-A-V-K, aa 45–48) which is required for kinase activity. This Lys is most likely involved in the phosphotransfer reaction, possibly functioning in proton transfer (Russo et al. 1985). The conserved aa Val is found as in other kinases also in *G. cydonium* two positions towards the C-terminus from the Gly-Xxx-Gly-Xxx-Gly, and is thought to stabilize the conserved glycines (Hanks et al. 1988). Asp (aa 51) and Asn (aa 56; both in subdomain VI) as well as the D-F-G trimer Asp (aa 169; subdomain VII), Phe (aa 170) and Gly (aa 171) is present in the sponge sequence; it has been implicated in ATP binding (Hanks et al. 1988). This triplet is the most conserved portion in the catalytic domains and is surrounded for two positions at both ends by the hydrophobic or near-neutral residues Ile and Ser (5′-position) and Met and Ser (3′-position). Asp $[D]^{151}$ in subdomain VIb, as well as Asp $[D]^{169}$ in subdomain VII, is thought to interact with the phosphate groups of ATP via a Mg^{2+} salt bridge (Brenner 1987).

The catalytic domain is terminated by subdomain XI which is characterized by the invariant Arg $(R)^{248}$, which is preceded by the Met $(M)^{237}$. The putative roles of the consensus sequences in subdomains III, IV, IX, X, and XI are not known; again, subdomain X is missing in the sponge RTK.

Tyr (aa 181; subdomain VII) undergoes phosphorylation (tyrosine kinase phosphorylation site) located seven residues to the C-terminal side of an Arg; this Tyr is located adjacent to the Tyr [aa 182], which is the potential autophosphorylation site.

Sequence alignments on the aa level of the sponge RTK with the conserved catalytic domains of the protein-tyrosine kinase family members were performed in a first step with an intracellular TK *Src* (from the freshwater sponge *S. lacustris*; Ottilie et al. 1992), human *Fes* and human *Abl* TKs (Beveren 1988), the integral membrane tyrosine kinase, the human insulin receptor (Ullrich et al. 1985) and human *Fer* (Sadowski et al. 1985; Hao et al. 1989) have been performed (Schäcke et al. 1994a). The two *src*-specific motifs Asp-Gly-Leu and Pro-Tyr-Pro-Gly-Met (D-G-L-P-Y-P-G-M) found in the deduced aa sequence of *S. lacustris src*-related genes (Ottilie et al. 1992) are not present in *G. cydonium* GCTK clone. The best homology was found with human insulin receptor, which belongs to class II of RTKs (according to Ullrich and

```
GCTK_1  ATCCGGGAAGTGAAACAGATTGGCGTCGGTCAGTTTGGAGCTGTGGTTCTTGCTGAAATGACTGGGCTCTCCGGGTCAGAACGTTGCGTC   90
GCTK_2  •••••••G•••••••••••••••••••••••••••••••••••••••••••••••••••••••••••••••••••••••A-•••••••••   89
GCTK_1   I  R  E  V  K  Q  I  G  V  G  Q  F  G  A  V  V  L  A  E  M  T  G  L  S  G  S  E  R  C  V    30
         •  •  •  •  •  •  •  •  •  •  •  •  •  •  •  •  •  •  •  •  •  •  •  •  •  •  N  V  A  S    30
        [-  -  -  -  -  -  -  -  -  -  I  -  -  -  -  -  -  -  -  -  -  -]

GCTK_1  CCTACCAAAAGGGATCCATCAATGCTGAACGGAGTAGCACTAGTGGCTGTGAAGAAACTAAAACCGGATGTGAGCGAGGAAGTGCGGCAG  180
GCTK_2  •••••••••••-•••••G•••••••T-••••••••••••••••••••••••••••••••••••••••••••••••••C•••••A•TT•••  177
GCTK_1   P  T  K  R  D  P  S  M  L  N  G  V  A  L  V  A  V  K  K  L  K  P  D  V  S  E  E  V  R  Q    60
GCTK_2   L  P  •  G  S  M  N  A  D  -  •  •  •  •  •  •  •  •  •  •  •  •  •  •  •  D  •  •  L  •    59
                                    [-  -  -  -  -  II -  -  -  -  -  -  -]       [-  -
                                                   #  #  #  #
GCTK_1  TCATTTGACAAGGAGATAAAGTTCGTGTCACAGCTCCAGCATGATAGCATCGTGCAGCTTTTGGCCGTATGTACACACAGTAAGCACCCC  270
GCTK_2  ••G••••••••A••••••••••••A•••••••A•••••••••••••••••••••••••••••••••A•••••••••••••••••••••••  267
GCTK_1   S  F  D  K  E  I  K  F  V  S  Q  L  Q  H  D  S  I  V  Q  L  L  A  V  C  T  H  S  K  H  P    90
GCTK_2   •  •  •  •  •  •  •  •  M  •  •  •  •  •  •  •  •  •  •  •  •  •  I  •  •  •  •  •  •  •    89
         -  -  III -  -  -  -  -  -  -  -  -]       [-  -  -  -  -  -  IV -  -  -  -  -  -]

GCTK_1  TTCATTGTGATGGAATACATGGAGAATGGAGACCTCAACCAGTTCCTGCAGAAATACCAGATGGTGGATGATGACTCTGCACTGTACTCA  360
GCTK_2  ••••••••••••••••••••••••••••••••••••••••••••••••••••••••••••••••••••••••••••••••••••••••••  357
GCTK_1   F  I  V  M  E  Y  M  E  N  G  D  L  N  Q  F  L  Q  K  Y  Q  M  V  D  D  D  S  A  L  Y  S   120
GCTK_2   •  •  •  •  •  •  •  •  •  •  •  •  •  •  •  •  •  •  •  •  •  •  •  •  •  •  •  •  •  •   119
        [-  -  -  -  -  -  -  -  -  -  -  -  -  -  V  -  -  -  -  -  -  -  -  -  -  -  -  -  -]

GCTK_1  AACCAGATCCCTCCCTCCACCCTCCTCTACATGGCCGTGCAGATCGCCAGTGGGATGGTGTACCTCTCCTCACTCAACTATGTCCACCGA  440
GCTK_2  ••••••••••••••••••••••••••••••••••••••••••••••••••••••••••••••••A•A••A•G••••••••••••••••••  447
GCTK_1   N  Q  I  P  P  S  T  L  L  Y  M  A  V  Q  I  A  S  G  M  V  Y  L  S  S  L  N  Y  V  H  R   150
GCTK_2   •  •  •  •  •  •  •  •  •  •  •  •  •  •  •  •  •  •  •  •  •  •  H  T  T  •  •  •  •  •   149
                                [-  -  -  -  -  -  -  VIa -  -  -  -  -  -  -]       [-  -  -  -  -
```

```
GCTK_1  GACCTCGCCACACGTAACTGTCTCGTCGGTTCCAACTTCCGTATCAAGATCTCCGACTTTGGCATGAGCCGCAACCTCTACGAGCGAGTC  540
GCTK_2  ••••••••••••••A••••••••••••••••••••••••••••••••••••••••••••••••••••AG••••••••••••••A••••••  537
GCTK_1  [D L A T R N] C L V G S N F R I K I S D F G M S R N L Y E R V  180
GCTK_2   • • • • • • • • • • • • • • • • • • • • • K • • • • • • • •  179
         - - - - VIb - - - - - - - -]     [- - - - - - - - - - VII - - -
         #         #                            # # #
GCTK_1  TATTACCGCGTCCGAGGCCGGGCAATGCTCCCCATTCGCTGGATGGCCACCGAGAGTTTCTACGGGCGATTCTCAGAGAAGTCAGATGCG  630
GCTK_2  ••••••••••••••••••••••••••••••••••••••T•••••••••••••••••C••••••••A••••••••••••••••••••••••  627
GCTK_1   Y Y R V R G R A M L [P I R W M A T E] S F Y G R F S E K S D A  210
GCTK_2   • • • • • • • • • • • • • • • • • • • • • • • • • • • • • •  209
         - - - -]  [- - - - - - - VIII- - - - - - -]   [- - - - - - IX -
         † ‡

GCTK_1  TGGGCGTACGGGGTGACTGTTTGGGAGATATACACGCTGGGGAAGAAGCAGCCGTACGAAGAGCTTGATGATCAGGATATGATCCAAGAT  720
GCTK_2  ••••••••••••••••••••••••••••••••••••••••••••••••••••••••T•••••••••••••••••••••••••••••••••  717
GCTK_1   W A Y G V T V W E I Y T L G K K Q P Y E E L D D Q D M I Q D  240
GCTK_2   • • • • • • • • • • • • • • • • • • • • • • • • • • • • • •  239
         - - - - - IX - - - - - - - - -]     [- - - - - - - - - XI - -

GCTK_1  GCCATTCGAGGCACGGGTCGTAGGATCATGGGTCGACCG-GGGGTGTCGCAGGCTGTGTATGAGGTGCTACTGCGGTGCTGGGAGTACGC  809
GCTK_2  •••••••••••••••••••••••••••••••••••••••A•••••••••••••••••••••C•••••••••••-••A••••••T••••••  806
GCTK_1   A I R G T G R R I M G R P G V S Q A V Y E V L L R C W E Y A  270
GCTK_2   • • • • • • • • • • • • • R G V A G C V R G A T • • • V • •  269
         - - - - - - - - - - - XI - - - - - - - - - - -]

GCTK_1  CGCCGCAGATCGAGCCACGTTCAAAGAGATACATGACAGCCTGAACCTCATTCAACTCAACAGC  879
GCTK_2  •••••••••••••••••••••T••G•••••••••••••••••••••••••••••••••••••••  876
GCTK_1   A A D R A T F K E I H D S L N L I Q L N S  293
GCTK_2   • • • • • • • • • • • • • • • • • • • • •  292
```

Fig. 9. Catalytic tyrosine kinase domain of sponge RTK. The two nucleotide sequences of sponge cDNA for RTK (GCTK_1 and GCTK_2) together with their deduced amino acid sequences are shown. The 11 subdomains are indicated. Identical nt as well as aa are marked by *filled circles*; gaps are indicated by *hyphens*. The core motif of the TK-specific signature is *boxed*; the indicator sequence for the TK catalytic domain is *boxed* and *stippled*; the ATP-binding site is *stippled only*. The aa required for kinase activity are marked #; the potential tyrosine phosphorylation sites † and the autophosphorylation sites ‡ are marked by *crosses*

Schlessinger 1990) or family III (according to Hunter et al. 1992). The identity of sponge RTK with human insulin receptor precursor was 31% with respect to the aa within the catalytic domain (Fig. 10). Class II of RTKs shows a specific sequence in the subdomain VII with the consensus Asp-Leu/Ile/Val-Tyr-Xxx_3-Tyr-Tyr-Arg [D-(LIV)-Y-x_3-Y-Y-R]. In the deduced aa sequence of *G. cydonium* RTK, the corresponding sequence is Asn-Leu-Tyr-Glu-Arg-Val-Tyr-Tyr-Arg [N-L-Y-E-R-V-Y-Y-R; aa 175–183]. Because of the potential equivalence of aa Asp and Asn (Dickerson and Geiss 1969), we suggest that the RTK from *G. cydonium* belongs to the class II TKs and propose the consensus Asp-Asn-Leu/Ile/Val-Tyr-Xxx_3-Tyr-Tyr-Arg[(DN)-(LIV)-Y-x_3-Y-Y-R] for those enzymes (Fig. 10).

Two tyrosine kinase phosphorylation sites (Arg or Lys positioned seven aa residues to the N-terminal side of the phosphorylated Tyr) are present in the juxtamembrane region of the RTK (Tyr^{116} and at Tyr^{143}) and one in the catalytic domain Tyr^{181}.

4.6 3′-Nontranslated Region

The nontranslated sequences of the two cDNA clones of the sponge RTK show pronounced strong differences in lengths; while GCTK_1 comprises only 160 nt, GCTK_2 has 258 nt in length (Fig. 11). Both sequences are terminated by a poly(A) stretch. Neither of the two sequences shows the typical signal polyadenylation site AATAAA usually present in higher eukaryotes (Zarkower et al. 1986). A truncated sequence (AAATAA) is located 23–19 nt upstream in GCTK_1 of the poly(A) segment. In previous studies analyzing sponge polyadenylation sites, a potential sequence TAG..TAGTGT was identified both for the lectin (Pfeifer et al. 1993b) and the ubiquitin gene (Pfeifer et al. 1993a; Müller et al. 1994a).

5 Proposed Function of the Sponge Receptor Protein-Tyrosine Kinase

Due to the simplicity of their cellular organization, Porifera (sponges) have been used for first studies of the mechanisms underlying specific cell adhesion in multicellular eukaryotic organisms (Wilson 1907). The following characteristics of the sponge system make it attractive as a suitable model to investigate basic mechanisms of cell-cell and cell-matrix interaction. (1) The loose and flexible embedding of the cells in the mesohyl compartment (Garrone 1978), (2) the high motility of the cells in the organism (Garrone 1978), (3) the relatively low specialization and high differentiation potency of the cells (Borojevic 1966), and (4) the availability of large amounts of relatively homogeneous starting cell material for biochemical investigations.

Already in Porifera three adhesion systems exist, a Ca^{2+}-dependent cell-cell adhesion system, a Ca^{2+}-independent cell-cell adhesion system, as well as a

```
                                                          Consensus pattern for RTK class II □□□□□□
GCTK_1    IREVKQIGVGQFGAVVLAEMTGLSGSERCVPTKRDPSMLNGVALVAVKKLKPDVSEEVRQSFDKEIKFVSQLQHDSIVQLLAVCTHSKHP   90
HSINSR    L--LREL•Q•S••M•YEGNARDIIKG•AETR-------------••••TVNESA•LRERIE•LN•ASVMKGFTCHHV•R••G•VSKGQPT 1098
                 * *                                    *                 *                 .

GCTK_1    FIVMEYMENGDLNQFLQKYQMVDDDSALYSNQIPPSTLLYMAVQIASGMVYLSSLNYVHRDLATRNCLVGSNFRIKISDFGMSRNLYERV  180
HSINSR    LV•••L•AH•••KSY•RSLRPEAENNPGRPPP-TLQEMIQ••AE••D••A••NAKKF••••••A•••M•AHD•TV••G••••T•DI••TD 1188
                                                                      *    *  *         ***

          □□□  Consensus pattern for RTK class II
GCTK_1    YYRVRGRAMLPIRWMATESFY-GRFSEKSDAWAYGVTVWEIYTLGKKQPYEEL-------------------------DDQDMIQDAIRG  244
HSINSR    •••KG•KGL••V•••AP••LKD•V•TTS••M•SF••VL•••TS•AE-•••QG•SNEQVLKFVMDGGYLDQPDNCPERVTDL•RMCWQFN 1276
                           *           *    *

GCTK_1    TGRRIMGRPGVSQA   258
HSINSR    PKM•PTFLEI•NLL  1290
             *
```

Fig. 10. Sequence alignment of the catalytic domain of sponge RTK GCTK_1 showing aa sequence identities with TK domains of human insulin receptor precursor (HSINR; Ullrich et al. 1986). Residues identical to those of GCTK are shown by *filled circles*; deletions are denoted by *hyphens*. The aa present in other serine/and threonine- and tyrosine-specific protein kinases are indicated by *stars* [*] under the sequences. The consensus pattern for RTKs class II is marked by *open squares*

```
GCTK_2    TTCGAGGCACGGGTCGTAGGATCATGGGTCGACCGAGGGGTGTCGCAGGCTGTGTACGAGGTGCTACTGCGATGCTGGGAGTACGCCGCC   90

GCTK_1                CGTTCAA-GAGATACATGACAGCCTGA-CCTCATTCAACTCAACAGCTGATGACTGTTTCACTTTCATTCAAGTTTT   75
GCTK_2    GCAGATCGAGCCA••••T••G••••••••••••••T•••A•••••••••••••••••••••••••••••••••••••----•••••••• 176

GCTK_1    ATAATGTATTTGTCATTCGTTCTGTGTCTGTCAGTGTTTGTGTGGTTAGTT-TATAAGTCAAAATAACAGGTTTTTTACACTGCTA_n  160
GCTK_2    •••••••••••---•-•••••••••••••••••••••••••••••••••••A•G••••••••••G••••••••••••••••••••A_n  258
```

Fig. 11. Nontranslated part of sponge GCTK_1 and GCTK_2 cDNA segment. Identical nt are marked by *filled circles;* gaps are indicated by *hyphens*. The truncated potential polyadenylation signal (AAAU/GAA) is *underlined*

cell-matrix adhesion system. The first consists of two main elements, the intercellular aggregation factor (AF) which binds to the cell surface-associated aggregation (AR). Hence, the cell-cell adhesion system is heterophilic and of the third order. After the discovery of the first AF in *Microciona prolifera* (Moscona 1963), a marine sponge, and the subsequent purification of the AF from *Geodia cydonium* (Müller and Zahn 1973) and *M. prolifera* (Henkart et al. 1973), a first molecular explanation of the operation of cell adhesion molecules became possible. The AF is a multiprotein complex with a sedimentation coefficient of 90S (Müller et al. 1983). In its "native form" the AF appears as a sphere with a diameter of 1000 Å, which displays a concave cup structure. After treatment of the AF with detergents, a core structure is obtained that appears as "sunbursts" with a circular center (diameter 1000 Å) and 25 radiating arms. Incubation for 1–3 days of isolated single cells with the AF, which functions species-specifically, causes a rearrangement of the cells first into aggregates and finally into a functional sponge. An intact AR has not yet been purified from Porifera. Under defined conditions, the terminal glucuronic acid of the AR is enzymatically removed, resulting in an inactivation of its ability to bind to the AF. The consequence of this enzymatic process is a cell separation. The deglucuronylation is a reversible event. This and related observations led to the formulation of the "modulation theory of cell adhesion" (Müller et al. 1979).

The main components of the cell-matrix adhesion system in Porifera are a lectin and its cell surface receptor. The lectin from the sponge *G. cydonium* is specific for D-galactose (Müller et al. 1983). The lectin is extracellularly localized and binds to a 67-kDa membrane-associated protein. The gene has recently been cloned (Pfeifer et al. 1993b).

In principle, the adhesion systems in other – higher – animal phyla are very similar to those of Porifera. In higher animals, a Ca^{2+}-independent cell-cell adhesion system exists besides the Ca^{2+}-dependent adhesion system; some of these have tyrosine kinase activity and are provided with Ig-like domains (Edelman and Crossin 1991; Geer et al. 1994). In previous studies, the presence of a similar system was proven also in sponge cells (Müller 1982). In order to establish this Ca^{2+}-independent cell-cell adhesion system also by molecular biological means, a corresponding gene, the RTK described here, was identified in *G. cydonium*. At present it remains unknown if the RTK is indeed involved in cell recognition or functions as a growth hormone.

In this context it should be mentioned that sponges are rich sources for bioactive low molecular weight compounds; among them are tyrosine kinase inhibitors (Sarma et al. 1993). The most prominent ones are aeroplysinin from *Verongia aerophoba* (Kreuter et al. 1992) and the recently described polyketide, halenaquinone, isolated from the marine sponge *Xestospongia carbonaria* (Lee et al. 1992). Also this finding supports the importance of tyrosine kinases in sponges.

6 Implication for Molecular Evolution of Metazoa

Until the past 20 years, evolutionary relationships among metazoan phyla could rely only on detailed descriptions of animal morphology and embryology. Phylogenetic relationships had been established, but often uncertainties remained because of difficulties in distiguishing between convergent or divergent features. Especially for deep branches in phylogenetic trees, additional and independent methods have to be applied. Therefore, the sequencing method of rRNA was used to resolve distant phylogenetic relationships among metazoan phyla. However, also this method had its limitations and failed in some approaches, especially in those relationships which included the lowest multicellular organisms, the phylum Porifera (Field et al. 1988).

Recently, we presented a further approach which relies on aa sequence data obtained from genes identified mainly from the marine sponge *G. cydonium* (Gamulin et al. 1994; Müller et al. 1994b; Müller 1995; Müller et al. 1995). We have selected such genes encoding proteins which are features of multicellularity. We presented evidence that two adhesion molecules/receptors from the sponge *G. cydonium*, the lectin and the receptor tyrosine kinase, described here in detail, display on the deduced aa level sequences which show high homology to the corresponding proteins from vertebrates. These data strongly support also on the molecular level the view that the kingdom Animalia is a monophyletic group of multicellular organisms.

A further question in the general phylogeny of Metazoa remains unsolved: that of the evolution of the selected adhesion and receptor molecules in the protist lineage. It is unlikely that these molecules evolved suddenly. Focusing on the sponge receptor tyrosine kinase, the structure consists of four domains (an extracellular immunoglobulin domain, a transmembrane domain, and two cytoplasmic domains) which indicates that this receptor is already highly complex, very likely composed of different building blocks during evolution. For the development of these elements in the eukaryotic protist lineage, a period of $\approx$1000 m.a. was necessary, according to present estimates.

7 Summary and Perspectives

In the present chapter we summarized the sequence data obtained from cloning of the sponge RTK. The analysis revealed that the gene is constructed in blocks, suggesting a blockwise evolution of this protein molecule in a manner proposed earlier (Kolchanov and Lim 1994). In addition, the data contribute to our earlier view that the kingdom Animalia (Metazoa) is of monophyletic origin. After having analyzed – in addition to the RTKs – other genes typical for multicellularity, e.g., those coding for adhesion molecules/receptors, e.g., lectin (Pfeifer et al. 1993b) and a nuclear receptor, a homeobox-like gene (Kruse et al. 1994), we present further evidence that Porifera should be placed into the kingdom Animalia. Therefore we propose a monophyletic origin of all animals.

Acknowledgments. We are grateful to Dr. Vera Gamulin (Institute Ruder Boskovic; Zagreb) for her advice. This work was supported by grants from the Deutsche Forschungsgemeinschaft, the Bundesministerium für Bildung, Wissenschaft, Forschung und Technologie (under the coordination of KFA-Jülich, Internationales Büro) and the Stiftung Volkswagenwerk.

References

Aroian RV, Koga M, Mendel JE, Ohshima Y, Sternberg PW (1993) The *let*-23 gene necessary for *Caenorhabiditis elegans* vulval inducation encodes a tyrosine kinase of the EGF receptor subfamily. Nature 348: 693–699

Barnes RD (1980) Invertebrate zoology, 4th edn. Saunders, Philadelphia

Beveren CV (1988) Overview of the tyrosine kinase oncogenes. In: Reddy EP, Skalka AM, Curran T (eds) The oncogene handbook. Elsevier, Amsterdam, pp 185–191

Borojevic R (1966) Etude experimentale de la différenciation des cellules de l'éponge au cours de son développement. Dev Biol 14: 130–153

Brenner S (1987) Phosphotransferase sequence homology. Nature 329: 21

Cooper JA, Esch FS, Taylor SS, Hunter T (1984) Phosphorylation sites in enolase and lactate dehydrogenase utilized by tyrosine protein kinases in vivo and in vitro. J Biol Chem 259: 7835–7841

Csaba G (1987) Why do hormone receptors arise? In: Csaba G (ed) Development of hormone receptors. Birkhäuser, Basel, pp 7–13

Csaba G (1994) Phylogeny and ontogeny of chemical signaling: origin and development of hormone receptors. Int Rev Cytol 155: 1–48

Cunningham BA, Hemperley JJ, Murray BA, Prediger EA, Brackenbury R and Edelman GM (1987) Neural cell adhesion molecule: structure, Ig-like domains, cell surface modulation and alternative RNA splicing. Science 236: 799–806

Dickerson RE, Geiss I (1969) The structure and action of proteins. Harper and Row, New York

Edelman GM, Crossin KL (1991) Cell adhesion molecules. Annu Rev Biochem 60: 155–190

Field KG, Olsen GJ, Lane DJ, Giovannoni SJ, Ghiselin MT, Raff EC, Pace NR, Raff RA (1988) Molecular phylogeny of the animal kingdom. Science 239: 748–753

Gamulin V, Rinkevich B, Schäcke H, Kruse M, Müller IM, Müller WEG (1994) Cell adhesion receptors and nuclear receptors are highly conserved from the lowest metazoa (marine sponges) to vertebrates. Biol Chem Hoppe-Seyler 375: 583–588

Garrone R (1978) Phylogenesis of connective tissue. Karger, Basel

Geer P, Hunter T, Lindberg RA (1994) Receptor protein-tryosine kinases and their signal transduction pathways. Annu Rev Cell Biol 10: 251–337

Glenney JRJ (1992) Tyrosine-phosphorylated proteins: mediators of signal transduction from the tyrosine kinases. Biochim Biophys Acta 1134: 113–127

Hanks SK, Quinn AM (1991) Protein kinase catalytic domain sequence database: identification of conserved features of primary structure and classification of family members. Methods Enzymol 200A: 38–62

Hanks SK, Quinn AM, Hunter T (1988) The protein kinase family: conserved features and deduced phylogeny of the catalytic domains. Science 241: 42–52

Hao QL, Heisterkamp N, Groffen J (1989) Isolation and sequence analysis of a novel human tyrosine kinase gene. Mol Cell Biol 9: 1587–1593

Harrelson AL, Goodman CS (1988) Growth cone guidance in insects: fasciclin II is a member of the immunoglobulin superfamily. Science 242: 700–708

Henkart P, Humphreys S, Humphreys T (1973) Characterization of sponge aggregation factor. A unique proteoglycan complex. Biochemistry 12: 3045–3050

Hunter T (1991) Protein kinase classification. Methods Enzymol 200A: 3–37

Hunter T, Lindberg RA, Middlemas DS, Tracy S, Geer P vd (1992) Receptor protein kinases and phosphatases. Cold Spring Harbor Symp Quant Biol 58: 25–41

Knoll AH (1994) Proterozoic and early Cambrian protists: evidence for accelerating evolutionary tempo. Proc Natl Acad Sci USA 91: 6743–6750

Köhidai J, Karsa J, Csaba G (1994) Effects of hormones on chemotaxis in *Tetrahymena*: investigations on receptor memory. Microbios 77: 75–85

Kolchanov NA, Lim HA (1994) Computer analysis of genetic macromolecules. World Scientific, Singapore

Kreuter MH, Robitzki A, Chang S, Steffen R, Michaelis M, Kljajic Z, Bachmann M, Schröder HC, Müller WEG (1992) Production of the cytostatic agent, aeroplysinin by the sponge *Verongia aerophoba* in in vitro culture. Comp Biochem Physiol 101C: 183–187

Kruse M, Mikoc A, Cetkovic H, Gamulin V, Rinkevich B, Müller IM, Müller WEG (1994) Molecular evidence for the presence of a developmental gene in the lowest animals: identification of a homeobox-like gene in the marine sponge *Geodia cydonium* Mech Ageing Dev 77: 43–54

Lee RH, Slate DI, Moretti R, Alvi KA, Crews P (1992) Marine sponge polyketide inhibitors of protein tyrosine kinase. Biochem Biophys Res Commun 184: 765–772

Livneh E, Glazer L, Segal D, Schlessinger J, Shilo BZ (1985) The *Drosophila* EGF receptor gene homolog: conservation of both hormone and kinase domains. Cell 40: 599–607

Marschalek R, Hofmann J, Schumann G, Bach M, Dingermann T (1993) Different organization of the tRNA-gene-associated repetitive element, DRE, in NC4-derived strains and in other wild-type *Dictyostelium discoideum* strains. Eur J Biochem 217: 627–631

Mehlhorn H (1989) Grudriß der Zoologie. Fisher, Stuttgart, pp 81–82

Morgan WR, Greenwald I (1993) Two novel transmembrane protein tyrosine kinases expressed during *Caenorhabiditis elegans* hypodermal development. Mol Cell Biol 13: 7133–7143

Moscona AA (1963) Studies on cell aggregation: demonstration of materials with selective cell-binding activity. Proc Natl Acad Sci USA 49: 742–747

Müller WEG (1982) Cell membranes in sponges. Int Rev Cytol 77: 129–181

Müller WEG (1995) Molecular phylogeny of Metazoa [animals]: monophyletic origin. Naturwissenschaften 82: 321–329

Müller WEG, Zahn RK (1973) Purification and characterization of a species- specific aggregation factor in sponges. Exp Cell Res 80: 95–104

Müller WEG, Zahn RK, Kurelec B, Müller I, Uhlenbruck G, Vaith P (1979) Aggregation of sponge cells; a novel mechanism of controlled intercellular adhesion, basing on the correlation between glycosyltransferases and glycosidases. J Biol Chem 254: 1280–1287

Müller WEG, Conrad J, Schröder C, Zahn RK, Kurelec B, Dreesbach K, Uhlenbruck G (1983) Characterization of the trimeric, self-recognizing *Geodia cydonium* lectin I. Eur J Biochem 133: 263–267

Müller WEG, Diehl-Seifert B, Gramzow M, Friese U, Renneisen K, Schröder HC (1988) Interrelation between extracellular adhesion proteins and extracellular matrix in reaggregation of dissociated sponge cells. Int Rev Cytol 111: 211–229

Müller WEG, Schröder HC, Müller IM, Gamulin V (1994a) Phylogenetic relationship of ubiquitin repeats of the polyubiquitin gene from the marine sponge *Geodia cydonium*. J Mol Evol 39: 369–377

Müller WEG, Schröder HC, Schäcke H, Müller IM, Gamulin V (1994b) Phylogenetic relationship of adhesion proteins and ubiquitin from the marine sponge *Geodia cydonium*. Endocytobiosis Cell Res 10: 185–204

Müller WEG, Müller IM, Rinkevich B, Gamulin V (1995) Molecular evolution: evidence for the monophyletic origin of multicellular animals. Naturwissenschaften 82: 36–38

Nishida Y, Hata M, Nishizuka Y, Rutter WJ, Ebina Y (1986) Cloning of a *Drosophila* cDNA encoding a polypeptide similar to the human insulin receptor precursor. Biochem Biophys Res Commun 141: 474–481

Ottilie S, Raulf F, Barnekow A, Hannig G, Schartl M (1992) Multiple *src*-related kinase genes, *skr*1–4, in the fresh water sponge *Spongilla lacustris*. Oncogene 7: 1625–1630

Pfeifer K, Frank W, Schröder HC, Gamulin V, Rinkevich B, Müller IM, Müller WEG (1993a) cDNA cloning of the polyubiquitin gene from the marine sponge *Geodia cydonium* which is preferentially expressed during reaggregation of cells. J Cell Sci 106: 545–554

Pfeifer K, Haasemann M, Ugarkovic D, Bretting H, Fahrenholz F, Müller WEG (1993b) S-type lectins occur also in invertebrates: unusual subunit composition and high conservation of the carbohydrate recognition domain in the lectin genes from the marine sponge *Geodia cydonium*. Glycobiol 3: 179–184

Russo WM, Lukas TJ, Cohen S, Staros VJ (1985) Identification of residues in the nucleotide binding site of the epidermal growth factor receptor-kinase. J Biol Chem 260: 5205–5208

Sadowski I, Stone JC, Pawson T (1985) A noncatalytic domain conserved among cytoplasmic protein-tyrosine kinases modifies the kinase function and transforming activity of Fujinami sarcoma virus $P130^{gag\text{-}fps}$. Mol Cell Biol 6: 4396–4408

Sarma AS, Daum T, Müller WEG (1993) Secondary metabolites from marine sponges. Ullstein-Mosby, Berlin

Schäcke H, Schröder HC, Gamulin V, Rinkevich B, Müller IM, Müller WEG (1994a) Molecular cloning of a receptor tyrosine kinase from the marine sponge *Geodia cydonium*: a new member of the receptor tyrosine kinase class II family in invertebrates. Molec Membrane Biol 11: 101–107

Schäcke H, Müller WEG, Gamulin V, Rinkevich B (1994b) The Ig superfamily includes members from the lowest invertebrates to the highest vertebrates. Immunol Today 15: 497–498

Schäcke H, Rinkevich B, Gamulin V, Müller IM, Müller WEG (1994c): Immunoglobulin-like domain is present in the extracellular part of the receptor tyrosine kinase from the marine sponge *Geodia cydonium*. J Molec Recognition 7: 272–276

Schlessinger J, Ullrich A (1992) Growth factor signaling by tyrosine kinases. Neuron 9: 383–391

Seeger MA, Haffley L, Kaufman TC (1988) Characterization of amalgam: a member of the immunoglobulin superfamily from *Drosophila*. Cell 55: 589–600

Shishido E, Higashijima S, Emori Y, Saigo K (1993) Two EGF-receptor homologues of *Drosophila*: one is expressed in mesodermal primordium in early embryos. Development 117: 751–761

Shoelson SE, Chatterjee S, Chaudhuri M, White MF (1992) YMXM motifs of IRS-1 define substrate specificity of the insulin receptor kinase. Proc Natl Acad Sci USA 89: 2027–2031

Stephens RM, Schneider TD (1992) Features of spliceosome evolution and function inferred from analysis of the information at human splice sites. J Mol Biol 228: 1124–1136

Stoddard BL, Biemann HP, Koshland DE (1992) Receptors and transmembrane signaling. Cold Spring Harbor Symp Quant Biol 54: 1–15

Sun H, Tonks NK (1994) The coordinated action of protein tyrosine phosphatases and kinases in cell signaling. Trends Biochem Sci 19: 480–485

Sun SC, Lindström I, Boman HG, Faye I, Schmidt O (1990) Hemolin: an insect-immune protein belonging to the immunoglobulin superfamily. Science 250: 1729–1732

Ullrich A, Schlessinger J (1990) Signal transduction by receptors with tyrosine kinase activity. Cell 61: 203–212

Ullrich A, Bell JR, Chen EY, Herrera R, Tetruzzelli LM, Dull TJ, Gray A, Coussens L, Liao YC, Tsubokawa M, Mason A, Seeburg PH, Grunfeld C, Rosen OM, Ramachandran J (1985) Human insulin receptor and its relationship to the tyrosine kinase family of oncogenes. Nature 313: 756–761

Ullrich A, Gray A, Tam AW, Yang-Feng T, Tsubokawa M, Collins C, Henzel W, LeBon T, Kathuria S, Chen E, Jacobs S, Francke U, Ramachandran J, Fujita-Yamagushi Y (1986) Insulin-like growth factor I receptor primary structure: comparison with insulin receptor suggests structural determinants that define functional specificity. EMBO J 5: 2503–2512

Van Hejine G (1987) Sequence analysis in molecular biology. Treasure Trove of Trivial Pursuit. Academic Press, London

Williams AF, Barclay AN (1988) The immunoglobulin superfamily – domains for cell surface recognition. Annu Rev Immunol 6: 381–405

Zarkower D, Stephenson P, Sheets M, Wickens M (1986) The AAUAAA sequence is required both for cleavage and for polyadenylation of Simian Virus 40 pre-mRNA in vitro. Mol Cell Biol 6: 2317–2323

Wilson HV (1907) On some phenomena of coalescence and regeneration in sponges. J Exp Zool 5: 245–258

Subject Index

Springer-Verlag and the Environment

We at Springer-Verlag firmly believe that an international science publisher has a special obligation to the environment, and our corporate policies consistently reflect this conviction.

We also expect our business partners – paper mills, printers, packaging manufacturers, etc. – to commit themselves to using environmentally friendly materials and production processes.

The paper in this book is made from low- or no-chlorine pulp and is acid free, in conformance with international standards for paper permanency.

MIX
Papier aus verantwortungsvollen Quellen
Paper from responsible sources
FSC® C105338

If you have any concerns about our products,
you can contact us on
ProductSafety@springernature.com

In case Publisher is established outside the EU,
the EU authorized representative is:
Springer Nature Customer Service Center GmbH
Europaplatz 3, 69115 Heidelberg, Germany

Printed by Libri Plureos GmbH
in Hamburg, Germany